EVOLVING CONCEPTS IN SEDIMENTOLOGY

Francis Pettijohn

EVOLVING CONCEPTS IN SEDIMENTOLOGY

Edited by
Robert N. Ginsburg

The Johns Hopkins University Press
Baltimore and London

Manufactured in the United States of America

The Johns Hopkins University Press, Baltimore, Maryland 21218
The Johns Hopkins University Press Ltd., London

Library of Congress Catalog Card Number 72–4016
ISBN 0–8018–1444–8

Library of Congress Cataloging in Publication data
will be found on the last printed page of this book.

FOR FRANCIS PETTIJOHN,

TEACHER AND GEOLOGIST

Contents

Preface ix

1. Mopping Up the Turbidite Mess
   Roger G. Walker 1

2. Experimental Geochemistry and the Sedimentary Environment:
   Van't Hoff's Study of Marine Evaporites
   Hans P. Eugster 38

3. The Odyssey of Geosyncline
   Kenneth J. Hsü 66

4. Concepts of Appalachian Basin Sedimentation
   Earle F. McBride 93

5. Carbonate Petrography in the Post-Sorbian Age
   Robert L. Folk 118

6. Biostratinomy: The Sedimentology of Biologically
   Standardized Particles
   Adolf Seilacher 159

7. Recent and Ancient Algal Stromatolites: Seventy Years
   of Pedagogic Cross-Pollination
   Paul Hoffman 178

The Johns Hopkins University Studies in Geology
No. 21, *Evolving Concepts in Sedimentology*

Preface

"What have we learned in sedimentology in the past seventy years?" We chose this question as the starting point for a collection of essays and a conference to honor the distinguished contributions and teaching of Francis Pettijohn, Professor of Geology at the Johns Hopkins University. The question seemed singularly appropriate, because it echoes Pettijohn's own critical attitude, and because the study of sediments and sedimentary rocks as a specialty has come of age in his lifetime.

As the compilation took shape, it became clear that seven essays could hardly be expected to chart seventy years of the development of a specialty that now has some 5,000 practitioners, major scientific societies and journals, special institutes and curricula in many universities, and an expanding role in both the academic and applied aspects of earth science. However, we could reconnoiter selected areas of this vast unknown terrain and try to outline their major elements. We hope that the results of these reconnaissances, the seven essays that follow, will give some ideas of how certain aspects and concepts of sedimentology developed, and that they may also serve to stimulate other working scientists to examine what we have found to be a fascinating and challenging aspect of our specialty.

It is a pleasure to acknowledge the help of several friends and colleagues in the organization of this book. I am indebted to Paul Potter, Raymond Siever, and Bob Nanz for advice and encouragement in developing the theme of the collection and in selecting the authors. I am grateful to Ernst Cloos, George Fisher, Owen Bricker, and Owen Phillips for arranging the most successful conference, at Johns Hopkins in January 1971, at which these essays were first read. I thank the Faculty of the Department of Earth and Planetary Sciences for subsidizing the publication of this book.

ROBERT N. GINSBURG

Miami, Florida
January 1972

AUTHORS

Roger G. Walker, Associate Professor of Geology,
McMaster University, Hamilton, Ontario

Hans P. Eugster, Professor of Geology,
The Johns Hopkins University, Baltimore

Kenneth J. Hsü, Professor of Geology,
Swiss Federal Institute of Technology, Zürich

Earle F. McBride, Professor of Geology,
The University of Texas at Austin

Robert L. Folk, Professor of Geology,
The University of Texas at Austin

Adolf Seilacher, Professor of Geology and Paleontology,
University of Tübingen, Tübingen

Paul Hoffman, Geologist,
The Geological Survey of Canada, Ottawa

EVOLVING CONCEPTS IN SEDIMENTOLOGY

Chapter One

Mopping Up the Turbidite Mess

ROGER G. WALKER

THE FASCINATION OF TURBIDITES AND TURBIDITY CURRENTS

The contributors to this volume in honor of Francis Pettijohn were asked to focus on major developments within the last seventy years. In the case of turbidites and turbidity currents, the history of the subject can be almost entirely written within the last thirty-five years. Indeed, we shall recognize that for geologists, the revolution in our thinking took place only in 1950. In that year, Kuenen and Migliorini published their classic paper "Turbidity Currents As a Cause of Graded Bedding." Our modern thinking about turbidites stems from that paper, and we may, therefore, claim that now, in 1971, the "turbidite" is twenty-one years old and has finally come of age.

Since 1950 about 1,000 papers have been written on turbidites and turbidity currents—a rate of about one each week. What is so fascinating about turbidity currents that inspires so many people to write so much about them? I can only offer my own reasons here, and these can be listed as size, power, velocity, mystery, and elegance.

Turbidity currents are truly one of the large-scale processes operative on the earth today. Data from known slumps which led to turbidity current flow, and data from the estimated size of ancient turbidite beds, indicate that volumes of sediment of the order of 10^6 to 10^{10} cubic meters have been displaced in single flows. When mixed with water, the total volume of a flowing turbidity current can reach 10^7 to 10^{11} cubic meters, or 0.01 to 100 cubic kilometers. The inferred deposits of individual turbidity currents reach thicknesses of several meters in the geologic record, and single beds have reportedly been traced for more than 100 km.

The second aspect concerns the power of the currents to erode and transport sediment. It has been postulated that many, if not all, submarine canyons have been cut by turbidity currents and that the sediment thus eroded has been deposited at the canyon mouth on huge submarine fans. The volumes of erosion in the Delgada and Monterey Canyons have been estimated as 300 and 450 cubic km, respectively, and the volumes of the associated depositional submarine fans are at least 100 times as great as the volume of the canyons (Menard 1964). Clearly, then, the turbidity currents have caused drastic modification of submarine topography, incising canyons in the slopes and blanketing the preexisting deep-sea topography with very flat aprons of sediment.

The third aspect of turbidity currents which has caught the imagination of oceanographers and geologists is their velocity. Although no direct measurements have been made, the data from several instances of telegraph cable breaks indicate much higher velocities than most flows on land. For example, the Grand Banks turbidity current appears to have traveled at an initial velocity of 30 km/hour—some scientists estimate this initial velocity at 40 or even 55 km/hour, depending on the exact interpretation of the source of the current. These velocities have caught the imagination, particularly because the deep oceans were previously believed to be stagnant, uneventful places.

The fourth fascinating aspect of turbidity currents is that no one has ever seen one in the oceans—and those which have been seen in reservoirs have been slow and very dilute. We are forced to postulate their existence in the deep ocean from indirect evidence and to assess their importance in the geological record from even more indirect evidence.

For the geologist, the turbidity current theory is simple and elegant. Since 1950 tens of thousands of graded beds stacked on top of one another have been interpreted as turbidity current deposits, and it can safely be stated that no similar volume of clastic rock can be interpreted so simply. Consider, for example, the complexity of adjacent environments and resulting lithofacies in deltaic situations, or the rapid facies changes associated with barrier bar/lagoonal environments.

The turbidite is elegant in its simplicity, yet presents a scientific challenge because the alleged depositional process has never been observed in the oceans and because of the conclusions that can be drawn regarding paleogeography, paleoslope orientation, depth of water, tectonism in the source area, and basin geometry. More recently it has been realized that the organization of turbidites in ancient basins may lead to an interpretation of the type of continental margin, and hence to the type of sea-floor spreading and lithospheric plate behavior (Mitchell & Reading 1969).

The main purpose of this essay is not a scientific review, nor a historical review. Recent scientific reviews include papers by Kuenen (1964, 1967), Menard (1964), Middleton (1970), and Walker (1970): some aspects of turbidites are covered in the book by Dzulynski and Walton (1965) on flysch and greywackes. There is no full historical review of the turbidity current theory, although the "petit historique du dogme des turbidites" by Jean Philippe Mangin (1964) makes amusing reading. Perhaps the objection to the turbidity current idea on the part of several French geologists is due to the fact that they cannot decide if the deposit is masculine or feminine (le, ou la "turbidite"?). As with hurricanes, the destructive force of the current surely demands "la turbidite."

Despite the fascination and elegance of the turbidity current theory, there is a much more important aspect, namely, that the theory represents (arguably) the *only* true revolution in thought in this century about clastic rocks. I will develop the idea of scientific revolution below, following the thesis of Thomas S. Kuhn (1970). The revolution in thought has affected our ideas of erosion, transport, dispersal, and deposition in modern ocean basins. It has also completely changed our interpretations of sandstones and conglomerates in ancient basins, with respect to depth of deposition, transport process, paleocurrents, and dispersal systems. I maintain that no other development in clastic sedimentology in this century has caused a complete change in thinking of comparable magnitude. Before proceeding to discuss turbidity currents and turbidites we must, therefore, look briefly at Kuhn's ideas of scientific revolution.

KUHN AND THE SCIENTIFIC REVOLUTION

Science has generally been regarded as cumulative. Generations of scientists have gradually sharpened and clarified our understanding of nature as part of gradual progress toward a complete understanding. Some of the steps toward the ultimate goal have been longer and more productive than others, but the edifice as a whole has progressed steadily. This cumulative aspect is one of the more important ways in which science differs from the humanities.

However, Thomas S. Kuhn proposed in 1962[1] that science does not proceed continuously and cumulatively, but by *scientific revolutions* (during which old knowledge and methods of working are rejected), followed by periods of *normal science*, during which the discipline is reconstructed, using some of the old knowledge, but much new knowledge also, and using new observations and techniques. A summary of Kuhn's ideas is given in Table 1.

During the early development of a subject, scientists tend to make random, noncumulative observations, and each scientist develops his own theory to account for the observations. At first none of these theories achieves general acceptance, but at some stage one of them proves particularly successful and is adopted as a "paradigm" for the development of the subject. Kuhn uses the term paradigm in two different senses (1970, p. 175); "On the one

---

[1] The second edition of Kuhn (1970) is quoted throughout, as it contains a postscript in which Kuhn answers some of his critics.

Table 1. Stages in scientific development (Kuhn 1970)

1. *Early random observations.* No guidance from preexisting theory: each worker develops his own hypotheses.
2. *Emergence of first paradigm.* One of the hypotheses proves successful and is adopted by a group of scientists—it then guides their research activities.
3. *Crisis.* Facts or experimental results are found to be at variance with the paradigm. As more and more discrepancies are found, a state of professional crisis may develop.
4. *Revolution.* A new theory, capable of explaining the discrepancies, emerges. During a scientific revolution, the old paradigm is rejected and replaced by a new one.
5. *Mopping up.* The new paradigm is elaborated during a period of "normal science" (or "mopping-up operations").

hand, it stands for the entire constellation of beliefs, values, techniques and so on shared by the members of a given community. On the other, it denotes one sort of element in that constellation, the concrete puzzle-solutions which, employed as models or examples, can replace explicit rules as a basis for the solution of the remaining puzzles of normal science." Hence, under the guidance of a paradigm, observations and experiments are no longer random, but are directed toward full elaboration of the paradigm. This is what Kuhn means by normal science, or "mopping-up operations." Indeed,

> mopping-up operations are what engage most scientists throughout their careers. They constitute . . . normal science. Closely examined, whether historically or in the contemporary laboratory, that enterprise seems an attempt to force nature into the preformed and relatively inflexible box that the paradigm supplies. No part of the aim of normal science is to call forth new sorts of phenomena; indeed those that will not fit the box are often not seen at all. Nor do scientists normally aim to invent new theories, and they are often intolerant of those invented by others. Instead, normal-scientific research is directed to the articulation of those phenomena and theories that the paradigm already supplies (Kuhn 1970, p. 24).

Although we may agree that scientists are often (or usually?) intolerant of theories invented by others, we cannot, at least in geology, accept Kuhn's statement that scientists do not normally aim to invent new theories. Because of the difficulty of experimental work in geology, new theories tend to blossom as new field observations become available.

As more and more phenomena are discovered and more experiments are performed, it is quite possible that results will begin to lead away from the paradigm. In an acute example, there will be a crisis period during which the existing paradigm is gradually discredited and finally, perhaps suddenly, rejected. A successful scientific revolution will then lead to the establishment of a new paradigm, followed by another relatively long period of normal science. A paradigm switch involves more than an unusually long and productive step in a philosophy of cumulative science. It implies a completely new way of looking at the world, and it lays down a new set of guidelines which a particular scientific community follows in its research. It involves rejection of the old paradigm and much of the knowledge and theory associated with it—and rejection is not part of a "cumulative" scientific philosophy.

Kuhn points out two other important aspects of scientific revolutions that we shall apply to turbidity currents. First, he notes (p. 93) that "as in political revolutions, so in paradigm choice—there is no standard higher than that of the relevant community." Also, he observes (p. 110) that "led by a new paradigm, scientists adopt new instruments and look in new places. Even more important, during revolutions scientists see new and different things when looking with familiar instruments in places they have looked before."

The examples of crisis and revolution that Kuhn quotes are those best known to most scientists, for example, the Copernican and Newtonian revolutions, X-rays, radioactivity, and relativity. The only geological example is Hutton's uniformitarianism. However, Kuhn does not discuss the special problems of geologists, which result from scale and time difficulties in experimentation, and the consequent adoption of the method of multiple working hypotheses. As we shall see, crises are difficult to recognize in geology because interpretive problems are commonly blurred by the multiplicity of hypotheses. Revolutions, however, can be recognized and, for the reasons stated earlier, the turbidity current hypothesis constitutes a revolution perhaps equal in magnitude and importance to the continental drift revolution. However, dating and defining the revolution is not simple. The turbidity current paradigm as we now understand it has four main roots, in oceanographic, reservoir, experimental, and geological studies. Within each of these disciplines there were minor crises and revolutions, typified by the contributions of Daly (oceanography), Kuenen and Bell (experimental), Bailey, Natland, and Migliorini (geology), and Grover and Howard (field studies of reservoirs). Although coming close to the modern turbidity current paradigm, these scientists did not have as much impact on the profession as Kuenen and Migliorini (1950), who pulled the four roots together and presented the turbidity current paradigm as a "viable package." The *impact* of the turbidity current revolution can therefore be dated at 1948–50, but crisis and revolution had been proceeding, hand-in-hand, during the 1930's and 1940's.

In this essay, I will examine how the turbidity current theory stands up as an example of a revolution and, also, how Kuhn's ideas of crisis and revolution stand up in the light of the historical development of the turbidity current theory.

EARLY, RANDOM OBSERVATIONS

Kuhn (p. 15) observed that "the road to a firm research consensus is extraordinarily arduous." Here, we will examine some of the roots of the turbidity current paradigm, stressing the information collected by oceanographers, geologists, and engineers concerned with density currents in reservoirs. Specifically, we will look for evidence of Kuhn's statement that "in the absence of a paradigm or some candidate for a paradigm, all of the facts that could possibly pertain to the development of a given science are likely to seem equally relevant. As a result, early fact-gathering is a far more nearly random activity than the one that subsequent scientific development makes familiar. Furthermore, in the absence of a reason for seeking some particular form of more recondite information, early fact-gathering is usually restricted to the wealth of data that lie ready at hand" (p. 15).

Lakes and Reservoirs

Observations in lakes in the 1880's follow exactly the pattern outlined above by Kuhn. Local residents around lakes Constance and Geneva knew that the muddy Rhine and Rhone river waters disappeared very rapidly beneath the clear blue lake waters. It was generally assumed that this sinking was due to the greater density of the river waters, and that this in turn was due to their lower temperature.

During the course of a hydrographic survey of the Swiss lakes (1883–85), Hörnlimann discovered deep channels in the bottoms of lakes Constance and Geneva, immediately in front of the points where the rivers flowed into the lakes. He did not connect the channels with the sinking river water, but in 1884 von Salis suggested that the cold river water plunged down and continued to flow on the lake floors, cutting the channels.

However, in 1885 Forel realized that the "channel" was simply a depression between two levees and was partly due to erosion and partly to deposition. This suggested to him that the sediment load of the rivers also helped to increase their density, and that their flow on the lake floor was controlled both by temperature and suspended sediment. In 1887 Forel expanded his earlier views, stressing the importance of sediment in causing the rivers to flow to the bottoms of the submarine channels. He revised his ideas about erosion and stated that deposition on the levees alone would be sufficient to form the intervening channel.

Other workers, notably Wey, Eberhard, Delebecque, Heim, Kleinschmidt, Collet, and Schmindle worked on the lakes and generally favored the temperature hypothesis. Their contributions have been reviewed by Johnson (1938, 1939).

These observations in lakes represent the early development of the turbidity current theory. Kuhn, with reference to the early, random fact-gathering stages of science, notes (p. 16) that "no natural history can be interpreted in the absence of at least some implicit body of intertwined theoretical and methodological belief that permits selection, evaluation and criticism." In our case, right from the start, the concept of density flow guided the interpretations, but it was not sufficiently well developed to permit selection or evaluation (temperature versus suspended sediment as a cause of the excess density).

A clearer evolution of ideas, from initial random observations to a clear statement of the problems involved, can be seen in the reservoir literature. One of the earliest observations was made by Lawson (1919), who noted that when the Rio Grande was exceptionally muddy, due to flash floods, the muddy water could flow for more than 30 miles along the floor of Elephant Butte Reservoir and pass through the outlet, still as a muddy stream. It apparently did not lose its identity by mixing with clean reservoir water, even over the distance of 30 miles. However, until about 1930 there was little systematic observation and measurement of density currents in reservoirs, although it was realized, with considerable consternation, that currents of this type were causing silting-up of many reservoirs.

After 1930 considerably more work was done, and in 1938 Grover and Howard published their classic account of density currents in Lake Mead. The paper stimulated eighteen written discussions which were published in the *Transactions of the American Society of Civil Engineers.* In their reply, Grover and Howard (1938, p. 782) noted that the discussions had disclosed "that there have been instances of flows of turbid water through reservoirs under a variety of conditions that cannot be evaluated because of lack of observational information; that there may be many instances of stratification of water in reservoirs due to differences in density, in many of which turbidity is not involved; and that there is an important and unexplored field of scientific research in relation to density currents and stratification in reservoirs which may have both scientific and practical significance." Although not specifically mentioned in this summary, it was well known that silt was being transported the length of Lake Mead by these currents—a fact of considerable concern to the reservoir engineers.

From these discussions, it is apparent that a scientific problem of considerable importance had been fairly clearly formulated, even to the stage of setting up a committee for the study of density currents (Interdivisional Committee of the National Research Council on Density Currents, formed in 1937). Although there was no real paradigm for density currents at that time, there was awareness of a phenomenon and some agreement as to which observations and measurements would be worthwhile to make. The importance of density currents as agents for transporting sediment was emphasized in a review by Bell (1942). He considered their origin, the relationship between driving force and velocity, their behavior, the size of transported particles, the effect of flocculation, and the quantity of sediment transported. In his conclusions, Bell (1942, p. 546) noted that "density currents are among the commonest of natural phenomena," and "by giving due consideration to the mechanism that drives density currents students may be able to clear up some of the baffling problems of sedimentation." However, the currents envisaged by Bell were of *low* density, and he questioned their effectiveness as agents of submarine erosion (1942, p. 513) "Ph. H. Kuenen performed laboratory experiments that he believed supported Daly's hypothesis [that of cutting submarine canyons by density currents]; but at present the consensus of opinion seems to be that the task is entirely too great for density currents alone to accomplish."

In summary, we can see that early random observations in lakes and reservoirs have led toward consensus regarding the phenomenon of density currents. At the same time, many of the outstanding problems in the field had been defined, and later research was guided by an over-all theory, so that observations and measurements could no longer be termed random. The crisis stage before the density current paradigm did not appear in the lake and reservoir studies, but, as we shall see later, it did appear in the geological and oceanographic literature.

Oceanography

Two aspects of oceanography contributed to the crisis which immediately preceded the turbidity current revolution. The first concerned the origin of submarine canyons, and the second concerned the deposition of sand on the abyssal sea floor.

Apparently, the first submarine canyon to be discovered was the Hudson, discussed by Dana in 1863. With continuing work by the U.S. Coast and Geodetic Survey, more and more canyons were discovered and surveyed, and several rival theories for their origin were proposed. The first theory suggested that the canyons were drowned river valleys, although diastrophism and erosion by submarine currents (Smith 1902) were also postulated. As long ago as 1897, Milne noted that the displacement of sediment along the canyon axis was a cause of telegraph cable breaking.

Although the work of Forel was known, the general opinion was that the small channels in lakes Constance and Geneva were exceptional, and that density currents alone could not account for the erosion of the enormous oceanic canyons.

Little thought was given to submarine canyons in the years 1905-30, but the adoption of echo-sounding equipment by the Coast and Geodetic Survey about 1928 gave a new impetus to canyon studies. As detailed charts began to appear, with contours extending down to about 5,000 feet below sea level, speculation about canyon origin began again. This speculation was based mainly upon the shape of the contour lines on the map, the geographic and geological setting of the canyons, and a small amount of additional information from bottom-grab samples. No current surveys within the canyons had been made, and in many areas the regional geology was too poorly known to be of any use in deciding between rival theories.

The impending crisis did not follow exactly along the lines outlined by Kuhn. In most disciplines, experimentation is relatively easy and is not beset by scale and time problems. Crises, according to Kuhn, stem from growing dissatisfaction with existing paradigms and from the failure of these paradigms to account for new observations. However, with submarine canyons we can hardly call guesses made from contour maps "theories," let alone paradigms. The *crisis* arose from the fact that none of the proposed multiple working hypotheses gained a wide following, and that each hypothesis violated basic concepts of geology held by supporters of rival hypotheses.

In the late 1930's, the leading hypotheses were:

1. Submarine erosion, by
   a) rip currents
   b) tsunamis
   c) spring sapping
   d) density currents
2. Subaerial stream erosion followed by drowning
3. Diastrophism

Objection to submarine erosion was based on the belief that currents could not cut enormous canyons in semi-consolidated rock, let alone granite (Carmel Canyon). Similarly, objection to subaerial stream cutting was based upon a large body of geological information that indicated that there was *not* a 5,000 foot lowering of sea level in the Pleistocene, coupled with the fact that many canyons do *not* pass landward into rivers.

Because of the multiplicity of working hypotheses, we can clearly see both the impending crisis *and* the emergence of the new paradigm in the late 1930's. Kuhn pointed out (1970, p. 86) that "often a new paradigm emerges, at least in embryo, before a crisis has developed far or been explicitly recognized." In our case, the paradigm emerged because of Daly's (1936) application of Forel's density current hypothesis to the erosion of oceanic canyons, and because of Kuenen's (1937) experimental testing of Daly's hypothesis.

In 1936 Daly published his classic paper on the origin of submarine canyons. He began by showing how unlikely were diastrophism and subaerial erosion as effective agents for canyon formation, and proceeded to his "new conception" of density currents. These were essentially sediment-charged currents which would flow downslope, seeking and enlarging any natural depressions in the sea floor. He believed that the sediment was originally thrown into suspension by storms and noted that, in the North Sea, sand is sometimes thrown up from "40 to 50 meters to the decks of laboring ships" (1936, p. 407). Daly commented (p. 410) that "on account of the difficulty of scale, convincing experiments with laboratory models are not to be readily performed" (a red rag to a bull, in the light of Kuenen's experiments the next year), and hence proceeded to calculate velocities "by analogy," using engineer's formulas, specifically, a Chezy-type equation. Daly was not sure what figure to use for density, and he did not quote specific velocities for density currents. It is apparent from the discussion, however, that he was thinking in terms of about 1 m/sec. Finally, Daly reviewed the work of Forel and others in lakes Constance and Mead, and concluded that the troubles of the density current hypothesis "seem incomparably less serious than those of the older explanations of submarine 'canyons' " (1936, p. 419).

Daly's ideas clearly stimulated Kuenen, who immediately began experimental work. In 1937 Kuenen (p. 330) noted that

> the hypothesis shortly brought forward by Daly is striking both in its simplicity and its originality. . . . (p. 331) no theory has yet been proposed by which one could explain a world-wide sinking of ocean level to the present position of the lower canyon ends. . . . we are therefore forced to seek a mode of formation that could operate below sea level. Normal currents are not sufficiently restricted, properly directed or swift enough to form gouges down the slopes. . . . . we are left with the choice between admitting the probability of Daly's hypothesis or owning that we are completely baffled. . . . only objections that appear quite insurmountable can turn the scales against the suspension current theory

—and finally, Kuenen notes that "several geologists who witnessed the experiments admitted being more favorably disposed towards the theory than at first" (p. 331).

Now that the oceanographic study of the origin of canyons had at least one viable working hypothesis, experiments could be planned to test aspects of the hypothesis. Specifically, Kuenen asked (1937, p. 331–32):

"1. Will a suspension flow down a slope under water?
2. Will this type of current continue to a considerable depth without losing its motive force in consequence of mixing with clear water?
3. Has such a current any erosive power?
4. Will such a current follow a slight initial gorge? For if not, the mechanism can never develop a gorge when starting to work on a comparatively smooth slope.
5. Is the rate of flow increased by enlarging the scale of the experiment? In other words can swifter currents be expected in nature than in the laboratory?"

Although Kuenen was at Gronigen University, the experiments were carried out at Leyden. Details of the experiments, with many pictures, are carefully described, and Kuenen shows that all five questions can be answered in the affirmative. Using Chezy-type equations, Kuenen calculated flow velocities between 0.4 and 3 m/sec for the velocity of the head. In this context, he quoted Schmidt's (1911) work on atmospheric cold fronts. Schmidt attempted to simulate these, using saline density currents flowing beneath water. He claimed that the velocity of the head of the flow was only half the velocity of the current following, although the experiments were performed in less than ideal conditions in the region between laminar and turbulent flow (Middleton 1970). However, it is surprising that Kuenen did not use Schmidt's result in his discussion of graded bedding (Kuenen and Migliorini 1950), because, clearly, if the velocity of the body of the current is greater than the head, sediment will constantly be refluxed into the head, and no longitudinal segregation of grain sizes will take place.

Kuenen concluded in 1937 by stating (p. 349) that "The fact that the estimates lead to quite reasonable results and no flat contradiction is encountered between available energy and work to be performed, may certainly help to allay fear that the density current hypothesis is unable to account for the size of the submarine canyons."

Shortly after Kuenen's experiments, Johnson (1938, 1939) coined the term *turbidity current* (1938, p. 235) for currents in which the excess density was due to suspended particulate matter, as opposed to "salinity currents" where the excess density was due to salinity. Johnson was a strong advocate of submarine spring sapping for the origin of canyons, a theory which has no supporters at all today.

Daly incorporated Kuenen's experimental work and restated the turbidity current theory in 1942. Kuenen himself resumed experimental work immediately following the war, and presented the results at the International Geological Congress in London in 1948 (published 1950a). As we shall see, this meeting was extremely important in exposing geologists to the *high-density* turbidity current theory and the idea that such currents could transport sand.

Surprisingly, neither Daly nor Kuenen gave much thought to the deposits of turbidity currents before the war. In 1937 Kuenen surmised (p. 344) that "as the suspension spreads out over a wide area on reaching the bottom of the slope, it will deposit its load on a much larger surface than a river arriving at base level. If the submarine gorges have been formed by subaerial rivers there should be delta like projections at the lower ends. But if on the other hand they were generated by submarine flow of a suspension, there need be only thin wide flanges of erosion products, if fine materials predominated over coarse particles rolled along the bottom." However, in 1937 there were only a few scattered reports of deep-sea sands, and both Daly and Kuenen envisaged their density currents as being mainly muddy, and hence not potentially capable of delivering much sand to the deep-sea floor. However, Kuenen does quote extracts from a letter written to him by Stetson, who suggests (in Kuenen 1937, p. 350) that "even if we cannot use these currents for cutting gorges, it seems to me that we have a very important mechanism which can be used for distributing sediment down the Continental Slope and out to the ocean basins."

At about this time (1936), a series of cores were taken on a traverse across the Atlantic, using the cable ship *Lord Kelvin*. The average length of these cores was a little over 2 m, and complete core descriptions were published by Bramlette and Bradley (1940). Shepard (1948, p. 292) remarked that "somewhat revolutionary information" resulted from the study of these cores, but did not discuss its significance. In fact, most of the cores contained sand layers, one with a sharp base and graded bedding (Bramlette & Bradley 1940, p. 15), which was explained (p. 16) as "material thrown into suspension by a submarine slump, carried beyond the slide itself, and deposited rapidly."

Bramlette and Bradley also commented on another anomalous feature which later proved to be very significant. The forams associated with the sandy layers represent warm-water forms, whereas those in the clays above and below the sand indicate cold-water forms. Bramlette and Bradley (1940, p. 16) suggested that the warm-water forms had been resedimented with the sand, a conclusion that had earlier been reached by Natland (1933) and was again to be reached independently by Migliorini (1946).

It is clear from Shepard's first edition of *Submarine Geology* in 1948 that the importance of sand on the deep-sea floor had not been fully realized. He does not even mention sand in his classification of deep-sea sediments, despite the earlier work and obvious anomalies mentioned by Stetson and Smith (1938) and Bramlette and Bradley (1940). However, in his book *Marine Geology*, Kuenen (1950b) seemed much more aware of the problem of deep-sea sands. He noted (p. 360) that sands had been "encountered especially in deep depressions of the Atlantic (north of Bermuda, Romanche Deep, Cape Trough, along the Whale Ridge, etc.), and in the basins west of southern California," and suggested that turbidity currents had been responsible for their emplacement. He discussed graded bedding briefly, referring to the "graded bedded facies" of E. B. Bailey (1936), and suggested that many beds in the fossil record might have been emplaced by turbidity currents.

A big impetus for oceanographic studies came with the development of piston coring devices in the late 1940's. Ericson, Ewing, and Heezen (1951) describe cores taken in 1947 and 1948 in which many sand layers (one more than 1 meter thick) were discovered, interbedded with clays and globigerina ooze. This stimulated a coring program of the flat plain of sediment beyond the entrenched Hudson Canyon, the cores being taken on the *Atlantis* expedition in 1950. The cores from this expedition showed many graded sand layers, with displaced shallow-water (Continental Shelf) forams in the coarse layers. Ericson, Ewing and Heezen concluded (1951, p. 964) that "transportation by turbidity currents most satisfactorily explains the distribution of sands in the sediments of the plain as well as the occurrence of gravel on the canyon floor." It is not certain what they meant by turbidity currents, because they do not quote Daly, Kuenen, or Johnson, and the term was not in common usage in the late 1940's. However, they do convincingly demonstrate erosion of the Hudson Canyon by the currents, considering this to be "proven by inclusion in the gravel of the upper Eocene chalk pebbles and cobbles of green clay from the canyon walls." The *Atlantis* cores were discussed more fully by the same authors in 1952, again not quoting the work of Kuenen.

The turbidity current theory arrived at a perfect time to avert a major crisis over the interpretation of the deep-sea sands. The theory neatly accounted both for the cutting of submarine canyons and the associated deposition of sediment on abyssal plains at the foot of the canyons. There was some opposition to the theory, particularly after publication of the 55-knot velocity calculated by Heezen and Ewing in 1952—but we will consider the opposition after looking at the evidence in the rocks.

Before continuing to rocks, however, we should pause and consider whether Kuhn's thesis of crisis, revolution, and normal science is in fact applicable to the oceanographic work, or whether the work on canyons and deep-sea sands was simply part of the cumulative scientific process that Kuhn argues against. The argument hinges upon whether one paradigm is simply being continuously elaborated, or whether there is a genuine conflict. For a crisis and revolution, Kuhn states (1970, p. 97) that "obviously, then, there must be a conflict between the paradigm which discloses anomaly and the one that later renders the anomaly law-like." It is suggested here that the old paradigm of oceanography disclosed a conflict when abundant graded sand and gravel layers were discovered on the deep-sea floor, because the old paradigm indicated that sand could only be transported in shallow, wave-agitated environments, that is, the Continental Shelf of pre-1950 oceanographers. The turbidity current paradigm rendered deep-sea sands law-like.

GRADED BEDS, AND A PALEONTOLOGICAL CRISIS

In the previous section, reviewing the development of lake, reservoir, and oceanographic studies, it has been shown that the turbidity current theory arrived at a very opportune moment, resolving the developing crisis in oceanographic studies of canyons and deep-sea sands. The theory was also presented at a very opportune moment for geologists, and helped to avert a crisis which undoubtedly would have developed with regard to the occurrence of "deep-water" forams associated with sandstones and conglomerates in the stratigraphic record. We shall therefore examine the early random geological observations of beds that we would now call "turbidites" (that is, the deposits of turbidity currents), and trace the pre-1950 studies to their state of impending crisis.

The first important observations were made by James Hall in 1843. He described features we would now call flute- and groove-casts from Upper Devonian rocks; and postulated that the grooves were made by objects larger than sand grains being transported on the bottom by ocean currents. He also observed that the orientation of the grooves, east to west, was consistent over several widely spaced outcrops. Clarke (1917), looking at the same rocks, called the sole marks "strand and undertow markings," and realized that the flutes had their deep noses upstream and that the marks had been gouged out by currents.

Elaborate classifications of sole markings were given by Vassoevich (1948, 1951), who worked on flysch-type rocks in the 1930's, 40's, and 50's. The basis of his classification was time of formation, *synglyphs* being formed during the deposition of the bed. The supposed mechanism of formation allowed synglyphs to be subdivided into *turboglyphs* (flutes), *proglyphs* (grooves), *bioglyphs* (organic markings), and *olistoglyphs* (marks formed by sliding at the sand-mud interface). This classification contrasts with the phlegmatic "ropy sole markings" of the British Geological Survey! Fortunately, we do not use Vassoevich's cumbersome terminology anymore, and most of the modern terms are purely descriptive—although they are grouped into two genetic types—tool marks and scour marks.

As well as displaying sole marks very nicely, the Upper Devonian of New York State also contains well-exposed internal sedimentary structures. These were noticed as early as 1928 by Pearl Sheldon, who wrote (p. 248) that

> . . . it appears that the sandstone beds were formed as units, and that the deposition of each passed through a regular succession of stages. Not all the stages are present in every bed but the full series consists of massive sandstone at the base grading into flat sedimentation surfaces followed by minute crossbedding and ordinary ripple marks, or large scale crossbedding probably associated with large, irregular ripples. Small scour channels may occur in place of the crossbedding, especially at the tops of lenses. It is probable that these stages depended upon the velocity of the water, the depth or volume of water and the sediment load. . . . The sediment was thoroughly churned at first, then distinguishable channels and flat sedimentation surfaces developed and finally the current diminished enough so that ripples were formed. These ripples were large and irregular where the currents were strong and carried a heavy load. The tops of the beds leveled off as the currents diminished further. Above the current bedding clay was deposited.

It is clear that Sheldon has described what we now call the Bouma (1962) sequence of internal structures ("massive sandstone at the base, grading into flat sedimentation surfaces followed by minute cross-bedding and ordinary ripple marks . . . above . . . clay was deposited"). Kuenen, when he looked at these rocks near Ithaca, was aware of Sheldon's work, but did not realize the significance of her description. The sequence of structures was formalized by Bouma in 1962 and was interpreted in 1965 in exactly the same way as Sheldon (but using flow-regime terminology) by Harms and Fahnestock, and

Walker. The significance of Sheldon's description was pointed out to me by
G. V. Middleton in 1966.

These sole-mark and internal-structure observations can only be termed
random, and none of the observations was made within the context of an
over-all hypothesis. The first major advance in the study of "turbidites" was
the recognition that graded beds constituted a distinct facies, characterized
by an assemblage of sedimentary structures. E. B. Bailey (1930, p. 85) records that it took him "many years to realize that graded bedding and current
bedding are the distinguishing marks of two different sandstone facies, the
one facies as important as the other." He also observed parallel lamination
and convolute lamination within the graded facies, and specifically pointed
out the *absence* of current bedding. (It is now realized that Bailey here is
referring to medium- to large-scale cross-bedding, not ripple cross-lamination.)
Bailey's work had an important influence on Kuenen, who also stated in the
early 1950's that current bedding and ripple marks are also entirely absent
from the graded facies. By the mid- to late- 1950's Kuenen and most other
writers had accepted that the tail of the turbidity current could form some of
the cross-lamination, and hence Kuenen stressed the absence of medium-scale
cross-bedding.[2]

However, to return to the 1930's, we must note the excellent descriptions
of the "graded facies" by Pettijohn (especially 1936, 1943) and Signorini
(1936). Signorini was one of the first of a long line of very careful Italian
geologists working on the "graded facies," and he observed graded bedding,
different types of sole marks, convolute lamination, and the use of these
features as way-up criteria. Signorini was apparently unaware of similar work
in other countries, particularly Russia, where Vassoevich was making careful
observations of sole marks, internal structures and their sequence, and facies
distributions within flysch troughs (details in Walker 1970).

The 1936 paper by Pettijohn is particularly important because the study
of the "graded facies" was undertaken within the framework of an existing
paradigm (the glacial paradigm), and did not simply comprise random observations. Pettijohn's observations in the Archaean of northwest Ontario included careful measurements of sandstone and slate thicknesses in several
long sequences of graded couplets, with a view to an attempted varve correlation following the pattern of De Geer, and Antevs. It is clear from reading
Pettijohn's discussion that he realized that the varve theory did not fully
account for the features he had observed, particularly that the "varves" did
not gradually become thinner up-section (indicating retreat of the ice), that
associated tillites were absent, and that there were no dropstones within the

---

[2] Although admittedly rare, medium-scale cross-bedding is by no means absent in
turbidites, and several examples have now been reported in the literature. It is also
believed now that much of the cross-lamination in turbidite sequences could be formed
by permanent ocean currents, rather than turbidity currents.

"varves." Pettijohn's observations on the varves do not depart far enough from the old paradigm to justify thoughts of a crisis, in Kuhn's sense. It is important, however, to record the feeling of dissatisfaction with the existing paradigm, because mounting dissatisfaction leads to crisis. To continue the northwestern Ontario story, Pettijohn and the writer spent three weeks together in the field in 1968, reinterpreting the Archaean rocks as turbidites (Walker & Pettijohn 1971).

Because of the importance of graded bedding in turbidites, it would be interesting here to look at some of the earlier descriptions, and note some of the interpretations. E. B. Bailey (1930, p. 85) records that "he first recognized graded bedding—and its significance as a criterion of succession . . . in the Spring of 1906," and at about the same time, graded bedding was being used as a way-up criterion in northern Wisconsin, in work under the direction of Hotchkiss (Cox & Dake 1916).

In North America, observation of graded bedding was routine (Tanton 1926, 1930; Bell 1929; Belyea & Scott 1935; Pettijohn 1936, 1943; Douglas, Milner, & McLean 1938), but opinions differed as to its interpretation. Shrock (1948, p. 78) records such interpretations as "seasonal changes, rhythmic rise and fall of sea level, tidal variations, periodic earthquakes and seaquakes and other less probable phenomena," but it is clear that there is no consensus. All authors, however, followed the old geological paradigm which stated that sandstones and conglomerates were shallow-water deposits. For example, Douglas, Milner, and McLean (1938, p. 34) noted that "the rock in places is finely laminated and may show cross-bedding of a minute type, attributed to current ripple, indicating deposition in shallow water" (with respect to turbidites of the Meguma Group of Nova Scotia; see Schenk 1970). They suggested a seasonal control of the coarse and fine layers. In view of Bailey's broad classification into "graded" facies and "current bedded" facies, it is interesting to note that Douglas, Milner, and McLean (1938) describe cross-bedding in association with graded bedding. This was also noted by Merritt (1934), in the Precambrian Seine clastics of northwestern Ontario, who commented that the graded bedding and current bedding both give the same way-up indication. However, he also pointed out that the graded bedding and cross-bedding do not occur together in the same bed, but in closely contiguous beds. Some time later, Ksiazkiewicz (1947, pp. 149–52) quoted Bailey's "graded" and "current bedded" facies, but commented that in the Carpathian Flysch, "some of flysch beds possess also well marked current bedding, although most of flysch beds exhibit only graded bedding." However, Ksiazkiewicz had perceived that the current bedding was of an unusual type, and he states that "one must assume that bottom currents have been responsible for this type of [current] bedding. Very likely one should ascribe this current bedding to undertow regularly working on the sea bottom. Conditions under which sand bars are deposited cannot be accepted because the sediments of this kind show great irregularities in cross-bedding." Ksiazkie-wicz follows Barrell (1917) in his interpretation, believing that occasional storms winnowed the sandstones, the finer winnowed material being deposited on top of the sand after the storm.

At about the same time that Pettijohn was working on the very oldest rocks, M. L. Natland was looking at the youngest. Specifically, Natland compared foram assemblages in different depth zones off the coast of southern California with assemblages collected in Hall Canyon from Plio-Pleistocene sands and conglomerates (the forams occurring in thin interbedded shales). Natland (1933) demonstrated that the five zones recognized in the modern ocean (Table 2) were also present in the Plio-Pleistocene rocks. Natland therefore suggested that the lowest sands and conglomerates may have been deposited in depths measured in several thousands of feet of water. In 1933 there were very few records of sands at this depth in modern oceans and no records at all of conglomerates, and Natland made no written comment on the implications of his paleontological work. However, he did suggest that the conglomerates were deep-water deposits, a suggestion which was either ignored or disbelieved. He has subsequently explained some of the opposition to his ideas in an address on "Paleoecology and Turbidites" (Natland 1963).

Table 2. Depth zones defined by M. L. Natland (1933) for modern and Plio-Pleistocene forams

Long Beach to Santa Catalina Island— Modern forams	Hall Canyon— Plio-Pleistocene
Zone	
I 1-7 ft.	I ⎫ all five
II 14-125 ft.	II ⎪ modern zones
III 125-900 ft.	III ⎬ present, and
IV 900-about 6,500 ft.	IV ⎪ represented by
V about 6,500-about 8,340 ft.	V ⎭ sandstones, conglomerates, and thin brown claystones

More than ten years later, a study of Tertiary forams was made by Migliorini, who was apparently unaware of Natland's work and of the findings of Bramlette and Bradley (1940). Migliorini (1946, p. 88) recognized that "the macroforams in their original position are found in pelitic or very fine psammitic rocks deposited in still water, whereas those in a 'secondary' position are found in coarse-grained rocks whose deposition demands a high transport capacity." Migliorini went farther than Natland, and suggested sand movements by submarine landslides—he was apparently unaware of the ideas of Daly and Kuenen. He stated (1946, p. 49) that "on the ocean floor, however, these landslides will not be able to behave as on land, because the material once displaced will mix with water and form a highly charged muddy stream, which will flow on the bottom as far as the slope allows, or else until, on the way, the detrital load becomes deposited."

It is clear from the work of Natland and Migliorini that some geologists were becoming aware that shallow-water fossils, together with large volumes of relatively coarse sediment, could be swept down onto the deep, quiet sea floor. It can be claimed that this is one of the roots of the pre-turbidite crisis, in the sense of Kuhn, because old paradigms are implicitly rejected—for example, those which indicated a decrease in grain size away from the shoreline, and those which indicated the exclusively shallow-water origin of coarse sandstones and conglomerates. However, the impression one gains from Kuhn is that crises affect entire scientific communities (e.g., communities of sedimentologists, or communities of marine geologists). He writes (1970, pp. 67–68) that "the emergence of new theories is generally preceded by a period of pronounced professional insecurity. As one might expect, that insecurity is generated by the persistent failure of the puzzles of normal science to come out as they should." In the turbidity current example, the work of Natland and Migliorini was particularly perceptive with regard to defining anomaly and crisis, but it by no means caused "pronounced professional insecurity," even within the community of sedimentologists or paleontologists. In fact, the "profession" either did not believe Natland, or considered his results to be a "local anomaly."

The same analysis can be made of the "graded facies." Few of the people who had worked on the "graded facies," let alone the profession, had been perceptive enough to realize that here was something which did not fit into any current theory of geology. For example, in the 1920's and 1930's, the Precambrian Halifax Formation (now termed the Meguma Group) of Nova Scotia was known to consist of nothing but graded sandstones and slates, about 30,000 feet thick. Bell (1929), Belyea and Scott (1935), and Douglas, Milner, and McLean (1938) all mention the graded bedding, but all appeal to the recurrent cross-lamination as evidence of deposition in shallow water in a slowly subsiding geosyncline. Bell (1929, pp. 24–26), for example, argues for a shallow-marine origin, stating that the absence of conglomerates makes a fluvial origin unlikely. He does not mention the fact that 30,000 feet of shallow-water deposits pose a problem, partly by their thickness but more importantly by their lack of facies changes, either vertically or laterally, throughout this thickness. Belyea and Scott (1935), with respect to the same rocks, wrote that "the rhythmical banding is the result of the rapid deposition of sands from rivers in flood followed by the slower deposition of slits after the flood season is past. These have accumulated in a shallow sea, the bottom of which has been stirred by waves and currents causing the typical cross-bedding."

The profession as a whole had *not* perceived an anomaly with respect to the "graded facies," and only a handful of geologists considered them to be a real problem. Of these, Bailey clearly described the important differences between the graded and cross-bedded facies, and perceptively assigned them to deep and shallow water respectively. He hoped (1936, p. 1717) that he "had persuaded some of his audience to follow him down the geosynclinal slopes, in company with earth-shaken grit and sand, to greater depths than are reached by ordinary bottom currents." Pettijohn (1943) was also aware of the anomaly. Implicit in his writings is the conclusion that if the Precambrian graded beds, with a "poured-in" petrographic aspect, are geosynclinal in Bailey's sense, then the glacial varve theory for their origin is not applicable.

However, most of the profession appears to have considered graded beds with bland indifference. Geologists, who are perhaps more at home with the method of multiple working hypotheses than physicists or chemists, could tentatively explain the graded facies by one or other of the working hypotheses. In this climate of thought, crises do not become very apparent to the profession, even though they may be perceived by a very few unusually aware scientists. If we acknowledge that the turbidity current theory constitutes a real geological revolution, we must disagree with Kuhn that there is necessarily a period of "pronounced professional insecurity" before a revolution. To a large extent, professional insecurity comes as a *result* of the revolution. For example, many geologists were forced to consider carefully whether their old ideas (say of paleocurrent patterns and source of coarse sediment) would have to be discarded and another interpretation worked out under the guidance of the new theory: or, alternatively, whether the new theory had to be rejected in the light of their own previous ideas and experience. Herein lies professional insecurity.

1948: BEGINNING OF THE REVOLUTION

At the International Geological Congress, held in London in 1948, Kuenen read a paper on "turbidity currents of high density." The paper presented the results of his experiments and stressed the importance of high-density flows, with specific gravities of up to about 1.6. This was an important development, because for the first time it introduced geologists to the idea of high-density sand flows, rather than the low-density muddy suspensions of Forel, Daly, Bell, and the reservoir engineers. This is not a mere matter of scale—the change from low- to high-density currents is of more significance to geologists and oceanographers than the original suggestion by Daly that density currents be the favored working hypothesis for canyon cutting. In the experiments, Kuenen showed that a 50 cm/sec current of clear water could just roll a grain weighing 0.16 gr. A density current of the same velocity (density not stated) could roll a grain weighing 1 kg., and he stated (1950*a*, p. 49) that "the experiments amply confirm the remarkable competency of turbidity flows deduced theoretically." With regard to turbidity currents of high density in nature, Kuenen (1950*a*, p. 49) stated that "a turbidity current, to deserve that name, must show complete mixing and renewed deposition of the transported material. Some sorting at least should also result after coming to rest. *The deposit of such a flow is probably inconspicuous in a sedimentary series* (my italics). The sole might show current-ripple marking, huge boulders might be enclosed or left stranded, gullying might have preceded deposition. Perhaps field geologists will eventually discover such evidence of deposition from turbidity flows if they keep the possibility in mind."

Kuenen's main conclusions concerned the erosion of submarine canyons, but he reiterated the idea that "even if the cutting power of density currents is considered doubtful, they may still be held as important agents in cleaning out existing furrows and in distributing sediment on the sea floor, as Stetson and Smith (1938) first suggested. They might possibly explain the observed occurrence of sand layers encountered here and there in deep-sea deposits far from the shore. *Many occurrences of graded bedding may be due to deposition from turbidity currents of high density* (my italics) when these have spread out and come to rest on basin floors, each bed resulting from a separate flow."

Kuenen's comments show that he was beginning to think more about deposits, even though his main emphasis was still oceanographic. At the Congress, Kuenen first used the term "graded bedding," but it is not elaborated in the paper, nor does Kuenen suggest that graded bedding would be an important feature for geologists to look out for. In fact, he seemed unaware of the many descriptions of graded bedding in the geological literature.

The significance of Kuenen's contribution was immediately recognized, and geologists seized his ideas, because they clearly resolved the previous paleontological crisis. Migliorini (in discussion of Kuenen 1950*a*, p. 52) argued that "when a high density current ceased to move larger blocks, the finer material would pass down gentle slopes to the lowest enclosed depression and there give rise to well graded sediments." Migliorini subsequently took Kuenen on his first turbidite field trip, to the "macigno" of the Appennines; Kuenen in turn realized the significance of graded bedding and went back to Gronigen and performed more experiments. The results of Kuenen and Migliorini's collaboration were published in the major "manifesto" of the "revolution," *Turbidity Currents as a Cause of Graded Bedding*, in 1950.

The micropaleontological work of Natland, discussed previously, was generally not believed by North American geologists, who could not accept the idea of deep-water sands and conglomerates. Dr. Ian Campbell, then at the

California Institute of Technology, heard Kuenen's paper in London and immediately suggested to Natland that his problems were over. Natland in turn arranged for Kuenen to visit North America on a lecture tour, and, as chairman of the Society of Economic Paleontologists and Mineralogists' Research Committee, suggested that there be a symposium on "Turbidity Currents and the Transportation of Coarse Sediments to Deep Water" at the society's 1950 meeting in Chicago. At that symposium, Kuenen (1951) read a paper stressing the high density of turbidity currents and the delta-like submarine fans they might be expected to form at the foot of submarine canyons. He also briefly discussed the nature of the deposits, and raised "another question of importance, . . . whether the coarse lower part of a graded bed deposited by a turbidity current must always be graded" (1951, p. 29). It seems clear that Kuenen had forgotten about Schmidt's experiments, which seemed to imply (because the velocity of the head is only half the velocity of the body of the current) that sediment is constantly shoveled into the head from the body, allowing no size segregation within the current and hence no grading in the lower part of the graded bed (Walker, 1965).

However, a much more important paper was read at the Chicago symposium by Natland and Kuenen (1951), elaborating Natland's micropaleontological results in the light of turbidity current theory. Natland and Kuenen together visited the Ventura Basin early in 1950 and carefully described conglomerates and sandstones. They emphasized the importance of deepwater forams in the shales, and resedimented shallow-water forams in the coarser sediments. Having made their point, that the coarse layers were emplaced under several thousand feet of water, they discussed at length the mechanism of emplacement, stressing the *absence* of "shallow water" sedimentary structures in the sandstones. For example (1951, p. 86), "ripplemarks, scour channels, shallow water mollusks, large scale cross-bedding and other markers for shallow water conditions, are almost entirely absent in the sandstones." And (p. 87), "cross-bedding is extremely rare in the sandstones and, in so far as observed, never clearly developed. In the silty parts of intervening shales, small-scale wavy cross-bedding—such as is normally associated with current ripples—was observed a few times. These occurrences are so haphazard and rare that they cannot be attributed to normal bottom stirring. It is suggested that they were formed by dilute turbidity currents."

Viewed in historical perspective, Kuenen's two papers with micropaleontologists form the basis of the new paradigm. The Kuenen and Migliorini paper stressed graded bedding, compared flume and field observations, and explaining *why* there was a fundamental difference between the two sandstone facies of E. B. Bailey. The paper with Natland, containing solid paleoecological data on displaced and in situ faunas, emphasized the resedimented nature of the beds and the fact that sands and conglomerates *could* be transported into very deep water. The *fact* of sands and conglomerates in deep water cannot be disputed, although today there is much discussion of process, particularly with regard to very coarse sediment.

One other geologist was immediately stimulated by Kuenen's paper at London in 1948; J. L. Rich immediately went to Wales and examined the Aberystwyth Grits. He commented (Rich 1950, p. 722) on the evenness of the bedding, sparsity of fossils, the minute scale of the cross-bedding, and the "flow casts" on the bottoms of the coarser layers. He noticed (p. 722) that "these show a striking alignment in a direction approximately N 25 E not only in a single bed, but also in all beds that show them. . . . this condition certainly points to some common cause for the markings and for the alignment. Whatever that may have been, it evidently was acting in the same direction through a considerable period of time while many feet of rock were being deposited." Rich, quoting Maxson and Campbell (1935), recognized that the "flute casts" had their deep noses upstream, and suggested (p. 728) that "the markings . . . are best explained as products of density currents rendered abrasive with respect to the underlying mud by a heavy burden of silt." Later, without quoting Kuenen, Daly, or Johnson, Rich (1950, p. 735) stated that "while the current was strongest, it would be a scouring agent, fluting the surface of the clay [which had been] deposited during the preceding period of calm. When the turbidity current ceased to flow, the silt remaining in suspension must have settled to form the silt beds as we now know them." Finally, Rich noted the similarity of the Aberystwyth Grits to the "flysch" of European geologists.

SUCCESS OF THE REVOLUTION: THE IMPACT OF THE NEW PARADIGM

The beginning of the revolution, for geologists, is difficult to date: so is the end. However, by the early 1950's the turbidity current paradigm was strongly directing research, and mopping-up operations had begun. In this section I will examine the paradigm as it appeared to the profession in the early 1950's, to see which new vistas of study it opened up, which old anomalies were made law-like, and whether the new theory suffered strong and convincing opposition. The four main aspects of the paradigm which guided research were (1) description (both of sedimentary structures and petrography); (2) paleocurrent and basin analysis of turbidites; (3) experimental study of turbidity currents; and (4) turbidity currents in the modern oceans. These aspects will be examined in turn, briefly recapping what was known before the revolution and examining Kuhn's statement (1970, p. 111) that "led by a new paradigm, scientists adopt new instruments and look in new places. Even more important, during revolutions scientists see new and different things when looking with familiar instruments in places they have looked before."

I. Sedimentary structures and petrography

Many of the main sedimentary structures of turbidites had been observed before 1950, including the sole marks, graded bedding, parallel lamination, cross-lamination, convolute lamination, pull-apart structures, and flame structures. Bailey had noticed that these features, particularly the ubiquitous graded bedding, characterized a distinct and widespread facies.

Following the revolution, the sole marks attracted immediate attention, both from the practical point of view in indicating paleocurrent directions, and in their intrinsic interest and variety. The 1950's were a period of cataloging, culminating in the publication of photo-encyclopedias by Dzulynski (1963) and Dzulynski and Walton (1965). The early observations tended to be qualitative, and the only systematic quantitative measurements have been made recently by Sestini and Curcio (1965) and Pett and Walker (1971). The latter writers attempted, for the first time, to relate sole-mark types to the internal-bedding features. Experimental reproduction of natural sole marks began in 1963 (Dzulynski & Walton) and has been continued by Dzulynski and his colleagues. However, quantitative experimental work has only been attempted in the last few years, with a study of flute casts by Allen (1969). The work on sole marks has made a tremendous contribution to paleocurrent and basin analysis, but so far, has contributed little to our understanding of turbidity current hydraulics. The flute-cast work should give an important insight into flow *before* the main episode of deposition begins.

The internal sedimentary structures had all been noticed before 1950, but Sheldon was the only person to have observed their organization. In the early 1950's, many types of graded bedding were described, and parallel lamination and ripple cross-lamination were observed. Ripple cross-lamination was used in paleocurrent studies, but its unreliability was only realized about 1960. Several turbidite models were proposed, but none achieved general acceptance until the Bouma sequence in 1962. This sequence neatly summed up and organized the earlier observations, and it has subsequently greatly influenced research. The sequence [A, graded bedding → B, parallel lamination → C, cross-lamination] can be interpreted in terms of decreasing flow regime. When this is done, it is apparent that cross-bedding, formed from migrating dunes, should theoretically occur between the parallel- and cross-lamination. This theoretical prediction has led to subsequent observation of cross-bedding in the field, but the rarity of cross-bedding in turbidites has in turn modified the original interpretation of the Bouma sequence (references in Walker 1970). The contribution of the internal structures to the turbidity current concept has therefore been in flow-regime interpretation, and in turbidite-facies definition discussed below.

However, despite the Bouma sequence and its general flow-regime interpretation, we have no quantitative explanation for graded bedding and no satisfactory explanation for parallel lamination. The parallel lamination has generally been attributed to "plane bed phase, upper flow regime," but this seems unsatisfactory, particularly because much of the parallel lamination in turbidites consists of interlaminated silt and clay—this is *not* the size of material generally associated with deposition in the upper flow regime. In fact, in Ordovician turbidites of Gaspé, Quebec, parallel lamination (silt and clay) passes laterally both upstream and downstream into ripple-drift cross-lamination (Bhattacharjee 1970).

The other descriptive aspect of turbidites concerns their petrography. This part of the mopping-up operations was initiated by Pettijohn in 1950, when he wrote a short discussion of Kuenen and Migliorini's paper in the *Journal of Geology.* Pettijohn's paper was entitled "Turbidity Currents and Greywackes," and he emphasized that although greywackes were well known in the geological record, he had "tried without success to find a greywacke in the process of formation" (1950, p. 169). He noted that the "chaotic appearance under the microscope, with scattered angular sand grains in a prominent to dominant 'clay' matrix" showed that "they are not the products of normal sedimentation" (Pettijohn, 1950, p. 169), and proceeded to show how chaotic deposition of sand and mud from a turbidity current could account for the petrographic features of greywackes. In the years following 1950 the term "greywacke" has had a checkered history: it has (unforgivably) been used interchangeably with the term turbidite, and it has given rise to arguments between those who believe the clay matrix content to be the essential part of the definition, and those who believe a high proportion of unstable rock fragments to be essential, with or without clay. Fortunately, it is now realized that the term "greywacke" should be purely petrographic and descriptive, and that the term "turbidite" is a field term applied to a rock believed to be the deposit of a turbidity current. Some turbidites are greywackes—others are clean sands because there was little clay or few rock fragments in the original current. Nevertheless, the arguments have clearly been guided by the turbidity current paradigm, with some workers seeking to interpret the mineralogy and texture of greywackes in the light of turbidity current depositional processes, and others in the light of rapid weathering and tectonic instability of the source lands of many geosynclinal areas.

While discussing the descriptive features of turbidites, we can note that very little of the opposition to the turbidity current theory is based upon the sedimentary structures. Kingma's (1958) objections are insubstantial and are based upon a misconception of how a turbidity current is believed to deposit sediment. The finding of sole marks in facies other than turbidites is no criticism of the turbidity current theory, although it does make the recognition of ancient turbidites more difficult. The mineralogy of turbidites is more or less irrelevant to their transport process and depends more upon the proportions of sand and mud (or plastic beads!) in the original current. No successful criticism of the turbidity current theory can be based on mineralogy.

2. Paleocurrents and basin analysis

I am sure Francis Pettijohn will have no objections to my borrowing his title for this section. It aptly sums up many of the consequences of the turbidity current paradigm—consequences which have broad implications not only for turbidites, but for the whole of geology.

The usefulness of paleocurrent measurement was established in the literature before 1950, so that when the turbidity current theory was introduced, paleocurrent analysis was immediately attempted by Rich in 1950. He showed transport directions toward N 25 E for the Silurian Aberystwyth Grits of the Welsh coast. The first turbidite paleocurrent rose diagrams were published by Kopstein, a student of Kuenen, in his thesis on some Cambrian rock in Wales in 1954. Crowell (1955) published rose diagrams and a paleocurrent map of the Prealpine flysch of Switzerland, and after 1955, paleocurrent measurement became a routine part of most turbidite studies. However, the early results of such studies were far from routine.

In 1954 (personal communication), J. E. Sanders, a postdoctoral fellow working with Kuenen, noticed sole marks in the Kulm of Germany, indicating flow parallel to the fold axes, which in turn were parallel to the long axis of the basin. As Sanders pointed out (personal communication), this would surprise many geologists who had previously assumed direct transport of sand from marginal land masses (real or imaginary) laterally toward the axis of the basin. The idea of longitudinal transport was emphasized by Kuenen and Sanders (1956) and by Kuenen (1957, 1958), who quoted modern oceanographic examples of elongate basins with supply dominantly at one end (the Gulf of California, and the Adriatic Sea, for example).

The concept of longitudinal flow revolutionized the interpretation of several basins. For example, O. T. Jones criticized Kuenen's and Kopstein's paleocurrent work in Wales, stating with reference to the Aberystwyth Grits (1956, p. 329) that "the thickness of individual beds of grit and of the greywacke suite as a whole diminishes eastward and it is evident that the sediments were derived largely, if not wholly from the *west*" (my italics). However, in 1954, again with reference to the Aberystwyth Grits, Jones (1954, p. 331) had previously commented that "perhaps the most plausible [method of emplacement] is that of Professor Kuenen's density-current hypothesis if he can persuade himself that the sediments were derived from the *east* and not from the southwest" (my italics). It is a reflection of the old paradigm (thinning and thickening of sedimentary wedges as simple indicators of source direction) that Jones was vague as to whether the source was in the *east* or *west*: likewise, it is a reflection of the new and successful paradigm that the problem could easily be solved.

In the Appalachians, the recognition of longitudinal flow in the Norman-skill and Martinsburg Formations (McBride 1962) also revolutionized our ideas of sediment dispersal. It did not change our ideas of sediment source, because lateral flow from the east and southeast toward the axis of the trough was also identified by McBride, agreeing with earlier ideas of an east or southeast source—"Appalachia."

As well as allowing reinterpretations of source and dispersal, the turbidity current theory allowed reinterpretations of depth of deposition. For example, Ruedemann and Wilson (1936) had proposed an "abyssal" origin for the cherts in the Normanskill, an interpretation which fitted very uneasily with the conventional shallow interpretation of the sandstones. With the guidance of the turbidity current paradigm, the sandstones could easily be reinterpreted as deep-water deposits, hence solving Ruedemann and Wilson's problem. The classic example of depth, however, is the Ventura Basin, where the turbidity current theory provided a mechanism to explain the earlier conclusions of Natland (1933; and Natland & Kuenen 1951), namely, that the sandstones and conglomerates had been deposited in several thousand feet of water.

As the turbidity current paradigm became elaborated by the efforts of normal science, basin interpretations gradually became more sophisticated. As well as using paleocurrent evidence for basin configuration, it was gradually recognized that turbidites could be divided into different lithofacies, the thick, coarse sandstones indicating proximal environments, and the thinner, finer sandstones indicating distal environments. This was first emphasized by Wood and Smith (1959) for the Aberystwyth Grits, and later work has shown that the proximal facies commonly show lateral paleocurrent directions, and the distal facies commonly show axial directions (Dzulynski, Ksiazkiewicz, & Kuenen 1959; Marschalko 1964, 1968; Stauffer 1967). As a development of the Bouma sequence, the sedimentary structures of turbidites have been used in an attempt to define "proximality" in a quantitative way (Walker 1967, 1970).

The studies of ancient turbidite basins and the interpretation of lithofacies have been strongly influenced by oceanographic developments. For example, the huge submarine fans which occur at the foot of most modern canyons have been identified in the geological record (Sullwold 1960, 1961; Walker 1966*a*; and several later papers), and fillings of deep channels have also been described (for example, Martin 1963; Walker 1966*b*; Normark & Piper 1969; Stanley & Unrug 1972). The discovery of contour currents in the oceans (Heezen, Hollister, & Ruddiman 1966) has made geologists much more cautious in interpreting current ripple paleocurrent indicators in ancient turbidites, but has not seriously affected the interpretation of sole marks. The contour currents are discussed below.

The final aspect of basin analysis concerns a provocative comment of Potter and Pettijohn (1964, p. 241) that "we cannot even be sure, however, that there is but a single model for all turbidite basins." It seems clear now that there are several types—three in the Appalachians alone, as defined by Pettijohn's students and colleagues. The simplest type was described by McIver (1961, 1970), who showed that in the Upper Devonian all the turbidity currents flowed consistently westward away from the advancing clastic wedge. There was no tectonic deformation until long after the turbidites had been covered by other sediments and no turbidity current flow parallel to the present tectonic axis. A second basin type was described by McBride (1960, 1962), who worked on the Middle Ordovician Martinsburg Formation. He demonstrated paleocurrent flow both perpendicular and parallel to the present tectonic axis. In the easternmost area, tectonic activity immediately followed deposition of the turbidites, but over most of the outcrop area there is an unbroken stratigraphic sequence into Upper Ordovician nonmarine sediments.

The third basin type is that described by Fisher (1970), Brown (1970), and Rankin (1970) in highly metamorphosed late Precambrian(?) rocks of the Blue Ridge and Piedmont areas. Here there is some evidence of transport *away from* the craton (Brown 1970, p. 346; Rankin 1970, pp. 235–38) as well as toward the craton. The basin was the only one of the Appalachian types to suffer extreme tectonics, metamorphism and subsequent uplift in the core of a mountain belt, and may represent, in part, a deformed Continental Rise wedge of turbidites.

3. Experimental study of turbidity currents

Since Kuenen's original experiments, several other studies have been made of turbidity currents and of flows designed to investigate one or more turbidite features. In the context of this essay, all of this work must be considered as normal science, and none of the experimental studies has generated any opposition to the basic paradigm. The review by Middleton (1970) summarizes this work very well, but points to the fact that progress has been slow. An explanation of the hydraulics of turbidity currents was given only as recently as 1966 (Middleton 1966a, b, 1967), and quantitative experimental work on flute casts began two years later (Allen 1968). There is still no quantitative work on the origin of graded bedding. The best experimental work on internal sedimentary structures is in the engineering literature (Simons, Richardson, & Albertson 1961), but it has had a great influence on the interpretation of flow regimes in turbidites (Harms & Fahnestock 1965; Walker 1965, 1967) and on proximal and distal depositional environments

(Parea 1965; Walker 1967)   Although the basic features of turbidites have been "mopped up" by geologists in the field, a great deal of experimental work must still be done to explain the known features.

4. Turbidity currents in the modern oceans

There has been a very important feedback from oceanographic studies into geological studies of turbidites during the years since 1950. In this brief look at the oceanographic work, I will stress those ideas which have been influential in the study of ancient rocks.

Perhaps the most important work was the calculation of the Grand Banks turbidity current velocity from the time at which different telegraph cables were broken. The earthquake in 1929 caused the instantaneous breakage of several cables, and then the sequential breakage of other cables at times ranging from 59 minutes to 18 hrs. 21 mins. after the quake. By plotting cable distance from the 59-minute break against time of break, Heezen and Ewing (1952, p. 868) calculated velocities for the turbidity current ranging from 55 knots (28 m/sec) at the 59-minute cable to 12 knots (6 m/sec) at a cable about 300 nautical miles from the 59-minute cable, after flow of about 13 hours.

These velocities astonished geologists and oceanographers. Kuenen (1952, pp. 879–80) attempted to calculate the velocity before the 59-minute cable, assuming a point source (now known to be incorrect), and gave a figure of 78 knots (90 miles per hour, or 40 m/sec !!). Shepard (1954) objected to both Heezen and Ewing's, and Kuenen's, velocities as "unimaginably high," and proceeded to look for an alternative mechanism for the cable breaks. Later, Shepard (1963, p. 340) ignored the 59-minute break and tried to fit a striaght line to the distance/time data for the later cable breaks, deducing a mean velocity of 15 knots (7.5 m/sec). The velocity was also considered "excessive" by Pettersson (1954, p. 75), who noted that other scientists, "both geologists and oceanographers, are skeptical of the hypothesis of Ewing and Bruce C. Heezen."

As a result of geophysical surveys (Heezen & Drake 1964, p. 222), it is now known that the earthquake generated both an enormous slump (which broke the first set of cables instantaneously) and a turbidity current. The slumped mass is some 50 miles long, up to 1,200 feet thick, and is of undetermined width. The main turbidity current probably originated by the coalescing of several smaller currents, and hence velocities upstream of the 59-minute cable cannot reliably be calculated. However, it does seem likely that the velocity of the main flow was at least as high as 37 knots (19 m/sec) (Menard 1964, pp. 208–10). These high figures have been noted with interest by geologists, but they have not affected geological thinking too much, because deposition is not assumed to have taken place. Some figures have been calculated for the velocities associated with the initial depositional phases of turbidity currents; the figures are in the range of 1–2 m/sec.

The second important contribution of oceanography to geology has been the measurement of more and more bottom-current velocities. It is now well known that these can make ripples in medium sand, and cut small scours in the lee of obstacles. As a result of these measurements, geologists have considered, with more and more emphasis, the possibility that many ripples on turbidite beds were made by ocean-bottom currents. This theory has been used in situations where there is a marked divergence between sole-mark and current-ripple directions (reviewed by Walker 1970, pp. 223–26). Ocean-bottom currents have even been suggested, by the few workers who still do not accept the turbidity current paradigm, as the prime agents for graded-bed deposition. The arguments against this possibility are overwhelming and have been reviewed by Kuenen (1964).

As a consequence of the bottom-current velocity measurements, and photography of the sea floor, it was discovered that many ocean currents flow *parallel* to the contours and not downslope. The first example to be discovered was the contour current flowing along the Continental Rise off the east coast of North America (Heezen, Hollister, & Ruddiman 1966). Contour currents have also been suggested as agents of graded-bed deposition, particularly in geological situations where there appeared to be an anomalous paleoslope. For example, there are about three well-documented examples of slump-fold orientations which superficially appear to indicate a paleoslope *perpendicular* to that indicated by sole marks. These situations have been used to argue against the turbidity current theory, but the arguments fail for many reasons, as reviewed by Walker (1970, pp. 223–26). However, it is now clear that in studies of ancient turbidites, sole-mark current indicators must be clearly separated from cross-lamination current indicators. It may even be possible to reconstruct ancient paleoslope and contour-current directions, although no such claims have yet been made.

There is one more important contribution from oceanography that was made by the JOIDES drilling program. At the time of writing, the full importance of the drilling for turbidite studies cannot be evaluated. Unfortunately, the rotary coring procedure almost eliminates the sedimentary structures in the turbidite sands, and hence the coring is unlikely to shed much light on the relationship between turbidite sedimentary structures and specific depositional environments. The most impressive result so far is the documentation of very long distances of turbidity current flow over extremely flat abyssal plains. For example, in the Vema Fracture Zone of the Mid-Atlantic Ridge, there is about 1000 m of sediment above basement, of which the upper 600 m was deposited within the last 500,000 years. The graded beds of the Vema Fracture Zone core are interpreted as turbidites, and the presence of abundant plant remains suggests derivation from the Amazon. If correct, this implies flow for 500 km across the flat floor of the Demerara Abyssal Plain (Benson, Gerard and Hay 1969, p. 662).

A second example of long-distance flow can be found in the Gulf of Mexico, where Tertiary turbidites are composed of terrigenous and organogenic material. The organogenic component appears to come from the Campeche Bank, but the volcanic part of the terrigenous turbidites suggests derivation from the Miocene Mexican highlands. In this case, the turbidity currents must have flowed for a little under 500 km (Ewing et al. 1969, p. 168).

In order to complete the essay, we should look back to the beginning—to the ocean-floor topographic problems of canyons, fans, and abyssal plains. It is now almost universally agreed that the fans and the topographically smooth abyssal plains have been built up by deposition from turbidity currents (Normark 1970)—ancient analogs have already been discussed. The canyons present a slightly greater problem, because the only observed erosive processes in modern canyons do not involve turbidity currents. The work of Shepard, Dill, and their colleagues (Shepard & Dill 1966) has shown that sand flow and sand creep are important erosive processes in modern canyon heads, but it has not been demonstrated that these processes can operate on gentle slopes in the lower canyons. Granted that the fan and abyssal plain sediment is turbidite, and granted that most of the material eroded from the canyon has been deposited on the fans and abyssal plains, the most obvious process for canyon erosion is turbidity currents (Menard 1964).

1971: THE REVOLUTION COMES OF AGE

In the previous sections of this essay, the four roots of the turbidity current paradigm have been examined—reservoir and lake observations, oceanographic observations, experiments, and geological observations. It was shown that crises in oceanography and geology were recognized only by a few perceptive scientists and not by the profession as a whole. Major crises were avoided by the timely formulation of the turbidity current theory, which was presented to geologists in London in 1948, and in the classic paper by Kuenen and Migliorini in 1950. The theory received overwhelming acceptance by the geological profession, but the last twenty-one years of "mopping up" have naturally raised many problems which have not yet been solved. These do not invalidate the paradigm—indeed, the problems only became apparent because of the paradigm. The main reasons for the success of the paradigm in this respect have been its simplicity, flexibility, and its predictive power.

The exciting days of real controversy about turbidity currents are over—the concept has lost some of its mystery but none of its elegance. After twenty-one years of intensive study, we may look ahead briefly at the aspects which still need "mopping up." First, although we know a fair amount about sole marks, internal structures, and petrography, we know little about the extent of individual beds, or the way in which bundles of beds are related to basin geography. More work along these lines is clearly of economic as well as academic importance. Second, we know little about the hydrodynamics of turbidity currents during flow, and have no quantitative explanation for graded bedding, parallel lamination, and sole marks. Third, we have no general understanding of the different types of turbidite basins. I believe that the highly metamorphosed turbidites in Piedmont situations present the greatest challenge to the turbidite sedimentologist, and also represent a situation in which the sedimentologist can make a unique contribution to ancient "large-scale" geology. The Piedmont turbidites of the Appalachians have been partially interpreted as ancient continental rise deposits (Brown 1970, p. 346), and their sedimentology has been described by Fisher (1970). The distribution of the facies and the paleocurrent directions should enable geologists to make interpretations of the continental margin and the type of geosyncline in terms of the types now recognized—Atlantic, Andean, Island Arc, and Japan Sea (Mitchell & Reading 1969). The gradual changes with time in paleocurrent directions, facies distributions, and tectonic behavior should enable interpretations to be made of ancient lithosphere plate coupling and uncoupling. However, before this can be done, sedimentologists will have to learn techniques for looking through kyanites at the original clay minerals, and for looking through highly deformed metagreywackes at the original graded-bedding and sedimentary structures. Now that the turbidity current paradigm is influencing and guiding studies in "large-scale" geology, it can truly be said to have come of age.

ACKNOWLEDGMENTS

The application of Kuhn's ideas to the development of the turbidity current theory, has been greatly stimulated by discussions with G. V. Middleton, and with participants at the Pettijohn Sedimentology Conference. I am indebted to B. C. Heezen, J. F. Hubert, Ph. H. Kuenen, M. L. Natland, and J. E. Sanders for the effort they made to answer my queries on points of historical fact and interpretation, and to those who made abundant critical and helpful comments on the manuscript—L. M. Cline, R. H. Dott, Jr., J. C. Harms, and G. V. Middleton. The responsibility for the facts and interpretations rests solely on my shoulders. Last, but not least, I am indebted to the National Research Council of Canada and the Geological Survey of Canada for continuing financial support of my research on turbidites.

REFERENCES

Allen, J. R. L. 1968. Flute marks and flow separation. *Nature 219*, 602–4.

———. 1969, Erosional current marks of weakly cohesive mud beds. *Jour. Sedimentary Petrology 39*, 607–23.

Bailey, E. B. 1930. New light on sedimentation and tectonics. *Geol. Mag. 67*, 77–92.

———. 1936. Sedimentation in relation to tectonics. *Geol. Soc. America, Bull. 47*, 1713–26.

Barrell, J. 1917. Rhythms and the measurement of geologic time. *Geol. Soc. America Bull. 28*, 745–904.

Bell, H. S. 1942. Density currents as agents for transporting sediments. *Jour. Geology 50*, 512–47.

Bell, W. A. 1929. Horton–Windsor district, Nova Scotia. *Geol. Surv. Canada, Mem. 155*, 1–268.

Belyea, H. R., & Scott, A. W. 1935. Conditions of sedimentation of the Halifax Formation as observed in Point Pleasant Park [Halifax, N.S.]. *Nova Scotian Inst. Sci., Proc. 18*, 225–39.

Benson, W. E., Gerard, R. D., and Hay, W. W., 1969. Summary and conclusions (for Leg 4), *in Initial Reports of the Deep Sea Drilling Project*, volume IV. Washington: U.S. Govt. Printing Office, 659–73.

Bhattacharjee, S. B. 1970. Ripple-drift cross-lamination in turbidites of the Ordovician Cloridorme Formation, Gaspe, Quebec. M.Sc. thesis, McMaster Univ., Hamilton, Ontario, 167 pp.

Bouma, A. H., 1962. *Sedimentology of some flysch deposits*. Amsterdam: Elsevier, 168 pp.

Bramlette, M. N., & Bradley, W. H. 1940. Geology and biology of North Atlantic deep-sea cores. *U.S. Geol. Survey, Prof. Paper 196–A*, pp. 1–34.

Brown, W. R. 1970. Investigations of the sedimentary record in the Piedmont and Blue Ridge of Virginia, *in* Fisher, G. W., Pettijohn, F. J., Reed, J. C., Jr., and Weaver, K. N. (eds.), *Studies of Appalachian geology, Central and Southern*. New York: Wiley–Interscience, pp. 335–49.

Clarke, J. M. 1917. Strand and undertow markings of Upper Devonian time as indication of the prevailing climate. *New York State Mus. Bull. 196*, 199–210.

Cox, G. H., & Dake, C. L. 1916. Geological criteria for determining the structural position of sedimentary beds. *Univ. Missouri School of Mines, Bull., 2*, 1–59.

Crowell, J. C. 1955. Directional-current structures from the Prealpine flysch, Switzerland. *Geol. Soc. America, Bull. 66*, 1351–84.

Daly, R. A. 1936. Origin of submarine "canyons." *Am. Jour. Sci. 31*, 401–20.

———. 1942. *The floor of the ocean*. Chapel Hill: Univ. North Carolina Press, 177 pp.

Dana, J. D. 1863. *A manual of geology*. Philadelphia: American Book Co., 798 pp.

Douglas, G. V., Milner, R. L., & McLean, J. 1937. The deposition of the Halifax Series. *Nova Scotia Dept. Pub. Works, Ann. Rept. Mines, pt. 2*, pp. 34–45.

Dzulynski, S. 1963. Directional structures in flysch. *Studia Geol. Polon. 12*, 1–136.

Dzulynski, S., Ksiazkiewicz, M., & Kuenen, P. H. 1959. Turbidites in flysch of the Polish Carpathian Mountains. *Geol. Soc. America Bull. 70*, 1089–1118.

Dzulynski, S., & Walton, E.K. 1963. Experimental production of sole markings. *Edinburgh Geol. Soc., Trans. 19*, 279–305.

―――. 1965. *Sedimentary features of flysch and greywackes*. Amsterdam: Elsevier, 274 pp.

Ericson, D. B., Ewing, M., & Heezen, B. C. 1951. Deep sea sands and submarine canyons. *Geol. Soc. America, Bull. 62*, 961–65.

―――. 1952. Turbidity currents and sediments in North Atlantic. *Am. Assoc. Petroleum Geologists, Bull. 36*, 489–511.

Ewing, M., et al., 1969. Site 3, Leg 1, *in Initial Reports of the Deep Sea Drilling Project*, Volume 1. Washington: U.S. Govt. Printing Office, pp. 112–78.

Fisher, G. W. 1970. The metamorphosed sedimentary rocks along the Potomac River near Washington, D.C., *in* Fisher, G. W., Pettijohn, F. J., Reed, J. C., Jr., and Weaver, K. N. (eds.), *Studies of Appalachian geology, Central and Southern*. New York: Wiley–Interscience, pp. 299–315.

Forel, F. A. 1885. Les ravins sous-lacustres des fleuves glaciaires. *Acad. Sci. Paris, Ct. Rend. 101*, 725–28.

―――. 1887. Le ravin sous-lacustre du Rhône dans le lac Léman. *Bull. Soc. Vaud. Sci. Nat. 23*, 85–107.

Grover, N. C., & Howard, C. S. 1938. The passage of turbid water through Lake Mead. *Am. Soc. Civil Eng., Trans. 103*, 720–90.

Hall, J. 1843. *Geology of New York, Pt. 4: Survey of the Fourth Geological District*. Albany, New York: Caroll and Cook, 683 pp.

Harms, J. C., & Fahnestock, R. K. 1965. Stratification, bed forms and flow phenomena (with an example from the Rio Grande), *in* Middleton, Gerard V. (ed.), *Primary sedimentary structures and their hydrodynamic interpretation. Soc. Econ. Paleontologists Mineralogists Spec. Pub. 12*, 84–115.

Heezen, B. C., & Drake, C. L. 1964. Grand Banks slump. *Am. Assoc. Petroleum Geologists, Bull. 48*, 221–25.

Heezen, B. C., & Ewing, M. 1952. Turbidity currents and submarine slumps, and the 1929 Grand Banks earthquake. *Am. Jour. Sci. 250*, 849–73.

Heezen, B. C., Hollister, C. D., & Ruddiman, W. F. 1966. Shaping of the Continental Rise by deep geostrophic contour currents. *Science 152*, 502–8.

Johnson, D. 1938. The origin of submarine canyons. *Jour. Geomorphology 1*, 111–29, 230–43, 324–40; *2* (1939), 42–60, 133–58, 213–36.

―――. 1939. *The origin of submarine canyons*. New York: Columbia Univ. Press. 126 pp.

Jones, O. T. 1954. The characteristics of some Lower Paleozoic marine sediments. *Roy. Soc. London, Proc., Ser. A. 222*, 327–32.

_____. 1956. The geological evolution of Wales and the adjacent regions. *Geol. Soc. London, Quart. Jour. 111*, 323–51.

Kingma, J. T. 1958. The Tongaporutuan sedimentation in Central Hawke's Bay. *New Zealand Jour. Geol. Geophys. 1*, 1–30.

Kopstein, F. P. H. W. 1954. Graded bedding of the Harlech Dome. Ph.D. Thesis, University of Groningen, 97 pp.

Ksiazkiewicz, M. 1947. Current bedding in Carpathian flysch. *Rocz. Polsk. Tow. Geol. 17*, 137–52.

Kuenen, P. H. 1937. Experiments in connection with Daly's hypothesis on the formation of submarine canyons. *Leidse Geol. Meded. 8*, 327–35.

_____. 1950*a*. Turbidity currents of high density. *Intl. Geol. Congr., 18th, London, 1948; Rept., pt. 8*, pp. 44–52.

_____. 1950*b*. Marine geology. New York: Wiley, 568 pp.

_____. 1951. Properties of turbidity currents of high density, *in* Hough, J. L. (ed.), *Turbidity currents and the transportation of coarse sediments to deep water. Soc. Econ. Paleontologists Mineralogists Spec. Pub. 2*, 14–33.

_____. 1952. Estimated size of the Grand Banks turbidity current. *Amer. Jour. Sci. 250*, 874–84.

_____. 1957. Longitudinal filling of oblong sedimentary basins. *Verhandel. Koninkl. Ned. Geol. Mijnbouwk. Genoot., Geol. Ser. 18*, 189–95.

_____. 1958. Problems concerning the source and transportation of flysch sediments. *Geol. en Mijnbouw 20*, 329–39. ,

_____. 1964. Deep sea sands and ancient turbidites, *in* A. H. Bouma, and Brouwer, A. (eds.), *Turbidites*. Amsterdam: Elsevier, pp. 3–33.

_____. 1967. Emplacement of flysch-type sand beds. *Sedimentology 9*, 203–43.

Kuenen, P. H., & Migliorini, C. I. 1950. Turbidity currents as a cause of graded bedding. *Jour. Geol. 58*, 91–127.

Kuenen, P. H., & Sanders, J. E. 1956. Sedimentation phenomena in Kulm and Flözleeres greywackes, Sauerland and Oberharz, Germany. *Am. Jour. Sci. 254*, 649–71.

Kuhn, T. S. 1970. *The structure of scientific revolutions*. Chicago: Univ. Chicago Press, 210 pp.

Lawson, L. M. 1919. Movement of silt, Elephant Butte reservoir. *Reclamation Record 10*, 411.

Mangin, J. P. 1964. Petit historique du dogme des turbidites. *Ct. Rend., Soc. Géol. France*, pp. 51–52.

Marschalko, R. 1964. Sedimentary structures and paleocurrents in the marginal lithofacies of the central-Carpathian flysch, *in* Bouma, A. H., and Brouwer, A. (eds.), *Turbidites*. Amsterdam: Elsevier, pp. 106–26.

_____. 1968. Facies distributions, paleocurrents and paleotectonics of the Paleogene flysch of central-west Carpathians. *Geol. Zbornik, Geol. Carpathica 9*, 69–94.

Martin, B. D. 1963. Rosedale channel—evidence for late Miocene submarine erosion in Great Valley of California. *Am. Assoc. Petroleum Geologists Bull. 47* 441–56.

Maxson, J. H., & Campbell, I. 1935. Stream fluting and stream erosion. *Jour. Geol. 43*, 729–44.

McBride, E. F. 1960. Martinsburg flysch of the Central Appalachians. Ph.D. thesis, The Johns Hopkins University.

———. 1962. Flysch and associated beds of the Martinsburg Formation (Ordovician), Central Appalachians. *Jour. Sedimentary Petrology 32*, 39–91.

McIver, N. L. 1961. Upper Devonian marine sedimentation in the central Appalachians. Ph.D. thesis, The Johns Hopkins University.

———. 1970. Appalachian turbidites, *in* Fisher, G. W., Pettijohn, F. J., Reed, J. C., Jr., and Weaver, K. N. (eds.), *Studies of Appalachian geology: Central and Southern.* New York: Wiley–Interscience, pp. 69–81.

Menard, H. W. 1964. *Marine geology of the Pacific.* New York: McGraw–Hill, 271 pp.

Merritt, P. L. 1934. Seine–Coutchiching problem. *Geol. Soc. America Bull. 45*, 333–74.

Middleton, G. V. 1966a. Experiments on density and turbidity currents. I, Motion of the head. *Can. Jour. Earth Sci. 3*, 523–46.

———. 1966b. Experiments on density and turbidity currents. II, Uniform flow of density currents. *Can. Jour. Earth Sci. 3*, 627–37.

———. 1967. Experimental studies of density and turbidity currents. III, Deposition of sediment. *Can. Jour. Earth Sci. 4*, 475–505.

———. 1970. Experimental studies related to problems of flysch sedimentation, *in* Lajoie, J. (ed.), *Flysch sedimentology in North America. Geol. Assoc. Canada, Spec. Paper 7*, pp. 253–72.

Migliorini, C. I. 1946. L'eta del macigno dell'Appennino sulla sinistra del Serchio e considerazione sul rimaneggiamento dei macroforaminiferi. *Boll. Soc. Geol. Ital. 63* (1944): 75–90.

Milne, J. 1897. Sub-oceanic changes. *Geo. Jour. 10*, 129–46, 259–89.

Mitchell, A. H. G., & Reading, H. G. 1969. Continental margins, geosynclines and ocean floor spreading. *Jour. Geology 77*, 629–46.

Natland, M. L. 1933: Depth and temperature distribution of some Recent and fossil Foraminifera in the Southern California region. *Bull. Scripps Inst. Oceanog., La Jolla, Tech. Ser. 3*, 225–30.

———. 1963. Paleoecology and turbidites. *Jour. Paleontology 37*, 946–51.

Natland, M. L., & Kuenen, P. H. 1951. Sedimentary history of the Ventura Basin, California, and the action of turbidity currents, *in* Hough, J. L. (ed.), *Turbidity currents and the transportation of coarse sediments to deep water. Soc. Econ. Paleontologists Mineralogists Spec. Pub. 2*, pp. 76–107.

Normark, W. R. 1970. Growth patterns of deep sea fans. *Am. Assoc. Petroleum Geologists Bull. 54*, 2170–95.

Normark, W. R., and Piper, D. J. W. 1969. Deep-sea fan-valleys, past and present. *Geol. Soc. America Bull. 80*, 1859–66.

Parea, G. G. 1965. Evoluzione della parte settentrionale della geosynclinale appenninica dall'Albiano all'Eocene superiore. *Atti Mem. Acc. Naz. Sc. Lett. Arti Modena 7*, 5–97.

Pett, J. W., & Walker, R. G. 1971. Relationship of flute cast morphology to internal sedimentary structures in turbidites. *Jour. Sedimentary Petrology 41*, 114–28.

Pettersson, H. 1954. *The ocean floor.* New Haven: Yale Univ. Press, 179 pp.

Pettijohn, F. J. 1936. Early Precambrian varved slate in northwestern Ontario. *Geol. Soc. America Bull. 47*, 621–28.

_____. 1943. Archaean sedimentation. *Geol. Soc. America, Bull. 54*, 925–72.

_____. 1950. Turbidity currents and greywackes—a discussion. *Jour. Geology 58*, 169–71.

Potter, P. E., & Pettijohn, F. J. 1964. *Paleocurrents and basin analysis.* New York: Springer-Verlag, 296 pp.

Rankin, D. W. 1970. Stratigraphy and structure of Precambrian rocks in northwestern North Carolina, *in* Fisher, G. W., Pettijohn, F. J., Reed, J. C., Jr., and Weaver, K. N. (eds.), *Studies of Appalachian geology, Central and Southern.* New York: Wiley–Interscience, pp. 227–45.

Rich, J. L. 1950. Flow markings, groovings and intrastratal crumplings as criteria for recognition of slope deposits, with illustrations from Silurian rocks of Wales. *Am. Assoc. Petroleum Geologists Bull. 34*, 717–41.

Ruedemann, R., & Wilson, T. Y., 1936. Eastern New York Ordovician cherts. *Geol. Soc. America, Bull. 47*, 1535–86.

Schenk, P. E. 1970. Regional variation of the flysch-like Meguma Group (lower Paleozoic) of Nova Scotia, compared to Recent sedimentation off the Scotian shelf, *in* Lajoie, J. (ed.), *Flysch sedimentology in North America. Geol. Assoc. Canada, Spec. Pub. 7*, 127–53.

Schmidt, W. 1911. Zur mechanik der Böen. *Meteor Zeit. 28*, 355–62.

Sestini, G., & Curcio, M. 1965. Aspetti quantitativi delle impronte di fondo da corrente nelle torbiditi dell'Appennino tosco-emiliano. *Boll. Soc. Geol. Italiano 84*, 1–26.

Sheldon, P. G. 1928. Some sedimentation conditions in Middle Portage rocks. *Am. Jour. Sci. 15*, 243–52.

Shepard, F. P. 1948. *Submarine geology.* New York: Harper (1st edition), 348 pp.

_____. 1954. High velocity turbidity currents, a discussion. *Royal Soc. London, Proc., Ser. A. 222*, 323–26.

_____. 1963. *Submarine geology.* New York: Harper (2nd edition), 557 pp.

Shepard, F. P., & Dill, R. F. 1966. *Submarine canyons and other sea valleys.* Chicago: Rand-McNally, 381 pp.

Shrock, R. R. 1948. *Sequence in layered rocks.* New York: McGraw-Hill, 507 pp.

Signorini, R. 1936. Determinazione del senso di sedimentazione degli strati nelle formazioni arenacee dell'Appennino Settentrionale. *Boll. Soc. Geol. Italiana 55*, 259–65.

Simons, D. B., Richardson, E. V., & Albertson, M. L. 1961. Flume studies using medium sand (0.45 mm). *U.S. Geol. Survey, Water Supply Paper 1498-A*, 76 pp.

Smith, W. S. R. 1902. The submarine valleys of the California coast. *Science 15*, 670–72.

Stanley, D. J., & Unrug, R. 1972. Submarine channel deposits, fluxoturbidites and other indicators of slope and base of slope environments in ancient marine basins, *in* Rigby, J. K., and Hamblin, W. K. (eds.), *Recognition of ancient sedimentary environments. Soc. Econ. Paleontologists Mineralogists Spec. Pub. 16*, 287–340.

Stauffer, P. H. 1967. Grain flow deposits and their implications, Santa Ynez Mountains, California. *Jour. Sedimentary Petrology 37*, 487–508.

Stetson, H. C., & Smith, J. F. 1938. Behaviour of suspension currents and mud slides on the Continental Slope. *Am. Jour. Sci. 35*, 1–13.

Sullwold, H. H. 1960. Tarzana fan, deep submarine fan of Late Miocene age, Los Angeles County, Calif. *Am. Assoc. Petroleum Geologists, Bull. 44*, 433–57.

_____. 1961. Turbidites in oil exploration, *in* Peterson, J. A., and Osmond, J. C. (eds.), *Geometry of sandstone bodies*. Am. Assoc. Petroleum Geologists, Tulsa, 63–81.

Tanton, T. L. 1926. Recognition of the Coutchiching near Steeprock Lake, Ontario. *Royal Soc. Canada, Trans. 20*, 39–49.

_____. 1930. Determination of age-relations of folded rocks. *Geol. Mag. 67*, 73–76.

Vassoevich, N. B. 1948. Le flysch et les méthodes de son étude. Gostoptekhizdat, Leningrad, 216 p. (Translated into French by Bureau de recherches géologiques et minières.)

_____. 1951. Les conditions de la formation du flysch. Leningrad: Gostoptekhizdat, 240 pp. (Translated into French by Bureau de recherches géologiques et minières.)

Von Salis. A. 1884. Hydrotechnische notizen: II Die Tiefenmessungen im Bodensee. *Schweiz. Bauzeit. 3*, 127.

Walker, R. G. 1965. The origin and significance of the internal sedimentary structures of turbidites. *Yorkshire Geol. Soc., Proc. 35*, 1–32.

_____. 1966a. Shale Grit and Grindslow Shales: transition from turbidite to shallow water sediments in the Upper Carboniferous of northern England. *Jour. Sedimentary Petrology 36*, 90–114.

_____. 1966b. Deep channels in turbidite-bearing formations: *Am. Assoc. Petroleum Geologists, Bull. 50*, 1899–1917.

_____. 1967. Turbidite sedimentary structures and their relationship to proximal and distal depositional environments. *Jour. Sedimentary Petrology 37*, 25–43.

_____. 1970. Review of the geometry and facies organization of turbidites and turbidite-bearing basins, *in* Lajoie, J. (ed.), *Flysch sedimentology in North America. Geol. Assoc. Canada, Spec. Paper 7*, 219–51.

Walker, R. G. & Pettijohn, F. J. 1971. Archaean sedimentation: analysis of the Minnitaki basin, northwestern Ontario, Canada. *Geol. Soc. America, Bull. 82* 2099–2130.

Wood, A., & Smith, A. J. 1959. The sedimentation and sedimentary history of the Aberystwyth Grits (Upper Llandoverian). *Geol. Soc. London, Quart. Jour. 114*, 163–95.

Chapter Two

Experimental Geochemistry And the Sedimentary Environment: Van't Hoff's Study Of Marine Evaporites

HANS P. EUGSTER

INTRODUCTION

Sediments are the products of physical, chemical, and biological processes acting throughout geologic time. If we desire to understand these processes, we have three choices: to study old rocks, to observe processes active today, or to imitate them. I want to address myself in this essay to the imitation of chemical processes of importance to sedimentary rocks.

Obviously, this is too broad a topic, because chemical experiments on rocks and minerals are as old as the study of sediments and of chemistry itself. The systematic use of chemical experiments to elucidate sedimentary processes, on the other hand, is quite recent. It coincides with the birth of experimental petrology as we know it today, and it is represented by Van't Hoff's study of marine evaporite equilibria carried out between 1896 and 1908.

It should come as no surprise that marine evaporites were chosen for that study. Not only are they the most typical examples of chemical sediments, but during the latter part of the nineteenth century their economic significance grew rapidly. The impact of Van't Hoff's study is widely acknowledged in general, but little attention is usually paid to why it was initiated. How was it that Van't Hoff, a leading physical chemist, became entangled with a geologic problem, and what response can we detect among later geologists to his pioneering efforts? This is a more narrowly formulated question, but one which I believe will lead to valid generalizations. In pursuing it, we will also have to review the development of phase theory by Bakhuis Roozeboom and

*An abbreviated version of this chapter appeared in *Science 173* (1971): 481–89.

the founding of the Geophysical Laboratory of the Carnegie Institution of Washington. We will find that the founders of that laboratory were significantly influenced by Van't Hoff's views and successes, while in Germany only salt geologists seemed to have followed in his footsteps. The study of sedimentary rocks remained virtually unaffected by his work. V. M. Goldschmidt, on the other hand, preferred to rely on Van't Hoff rather than on Gibbs for his classic interpretation of the metamorphic rocks of the Kristiania region.

BACKGROUND

In order to understand the reasons for Van't Hoff's choice of marine evaporites as a principal research topic, we must briefly review his background and previous contributions. Van't Hoff was born in Rotterdam, the son of a doctor. He showed an early interest in the natural sciences and particularly in chemical experimentation. After some initial training as a practicing chemist, he decided to study mathematics at the University of Leiden and he eventually obtained a Ph.D. in physics and mathematics from the University of Utrecht (1874). In the meantime, his love for chemistry reasserted itself and he worked with Kekulé in Bonn and Wurtz in Paris, both centers of the new science of stereochemistry. A major publication, entitled *Chemistry in Space*, established his reputation (Van't Hoff 1874). Subsequently, he taught chemistry at Utrecht and wrote a book on organic chemistry. He had become interested in organic reactions and particularly their temporal progress, which led him to study chemical kinetics. In 1877 he was called to the new University of Amsterdam as a lecturer, and in 1878 was appointed professor of chemistry, mineralogy, and geology. In a short time he acquired a large circle of students and collaborators. Their work was concerned principally with chemical kinetics and a summary was published in 1884: *Studies in Chemical Dynamics* (Van't Hoff 1884). The book became immediately famous. In addition to defining the concept of reaction rate and measuring the effect of temperature on it, Van't Hoff was concerned with chemical equilibrium and enunciated, for the first time, his "law of the incompatibility of condensed systems." This law is a special case of the phase rule which had been published six years earlier by Gibbs but with which Van't Hoff was unfamiliar. Van't Hoff contemplated different types of equilibria and divided them into physical (melting, vaporization) and chemical equilibria (chemical reactions, phase transformations). Among the latter he concerned himself with homogeneous, heterogeneous, and condensed (no vapor phase present) cases and he was struck by the difference between equilibria such as

$$2\,H_2 + O_2 \rightleftharpoons 2\,H_2O \text{ and}$$

rhombic sulfur $\rightleftharpoons$ monoclinic sulfur

The first he called a mobile equilibrium whose position shifts with temperature, while he noted that the latter exhibits a *transition point* at which the two phases can coexist but above or below which the two phases are incompatible. He was led to this conclusion by published observations on ammonium nitrate (p. 140) and he stated very clearly (p. 143) that if monoclinic and rhombic sulfur were brought together in a vacuum, a vapor equilibrium would be established until the maximum sulfur pressure was attained. If that pressure was higher than the equilibrium vapor pressure of one of the solids, the vapor would condense on that solid and the solid with the higher vapor pressure would volatilize until it was exhausted.

The point of transition is the temperature at which the vapors of the two bodies have maximum and equal tensions[1] (p..143).

DOUBLE SALTS AND SALT EQUILIBRIA

As I shall show shortly, the stage was now set for Van't Hoff's interest in evaporites. During the next two years, however, his attention was focused in another direction. The last section in the *Studies in Chemical Dynamics* dealt with problems of chemical affinity. In order to measure affinity, he used semipermeable membranes and he derived the Van't Hoff equation for gases and dilute solutions:

$$\frac{d \ln K}{dT} = \frac{\Delta H}{RT^2}.$$

This work was published in Sweden in 1886 (Van't Hoff 1886a), and immediately was hailed as a major contribution. Even the remaining discrepancies in the theory were extremely important, as they led Arrhenius to propose his theory of electrolytic dissociation (Arrhenius 1887). International fame was now assured, and in 1887 Van't Hoff was called to a chair in Leipzig, where Wilhelm Ostwald, the leading German physical chemist, taught. He declined the offer, as Amsterdam had promised him a new Institute. Nevertheless, he became co-founder and co-editor, with Ostwald, of the new periodical *Zeitschrift für Physikalische Chemie*, in which much of his subsequent work was published.

Meanwhile, Van't Hoff continued to concern himself with chemical equilibria. He extended his concept of transition points and phase incompatibility to hydrated salts, such as:

$$Na_2SO_4 \cdot 10\,H_2O \rightleftharpoons Na_2SO_4 + 10\,H_2O$$
mirabilite $\rightleftharpoons$ thenardite,

[1] All quotes from the German translated by the author.

again comparing this "chemical" transformation to "physical" melting. The next step was obvious. What about more complex salts, such as double salts? Do they also have transition points? He asked one of his students, Van Deventer, to check a double salt and the choice of the particular salt was crucial: $Na_2Mg(SO_4)_2 \cdot 4 H_2O$, known as astrakanite or bloedite from the Stassfurt deposits. The results, fully supporting his "incompatibility law" were published in the first issue of the *Zeitschrift für Physikalische Chemie* (Van't Hoff & Van Deventer 1887). This paper is important as a model for much of the subsequent work on salt systems. It contains data for the reaction

$$Na_2SO_4 \cdot 10 H_2O + MgSO_4 \cdot 7 H_2O \rightleftharpoons Na_2Mg(SO_4)_2 \cdot 4 H_2O + 13 H_2O$$

   mirabilite   epsomite   bloedite

The transition temperature was located at $21.5°C$, using the volume change of the reaction as an indicator. Simple dilatometers, topped with oil to prevent evaporation, were used and the observations were found to be most consistent when a mixture of reactants and products was used as starting material. Van't Hoff and Van Deventer also were interested in the relationship between solubility curves and the transition temperature. They found that "the solution of the stable system is more dilute, while that of the metastable system exhibits the properties of a super-saturated solution and indeed crystallizes by contact with the components of the stable system. . . . at the transition temperature itself the solubility of the two systems is equal, in other words that temperature corresponds to the intersection of the two solubility curves." (p. 178). Accurate solubility determinations are difficult to make and hence they decided to "determine the intersection of the vapor pressure curves of the saturated solutions instead of that of the solubility curves; since the vapor pressure depends only on the concentration, the two intersections must be identical" (p. 180). In this manner, the transition point was again found to lie at $21.5°C$.

Van't Hoff and Van Deventer went one step further and determined the effect of additional salts on the transition temperature. In the presence of halite, they located it at $5°C$ and concluded that this "extends the analogy between melting point and transition temperature; indeed we have here a situation which is completely analogous to the depression of the melting point of ice in the presence of salts" (p. 182-83).

ORIGIN OF PHASE THEORY

The work of Van't Hoff and Van Deventer attracted the attention of another Dutch chemist, Bakhuis Roozeboom, who was later to be Van't Hoff's successor as professor of chemistry in Amsterdam in 1896. At the time, Roozeboom was working in Ostwald's laboratory in Leipzig and he had just published (Roozeboom 1888*a*) an extensive summary of his work (1884-88) on chemical equilibria, in which he classified different types according to Gibbs's phase rule. Van der Waals, then professor of physics in Amsterdam, had made him aware of Gibbs's contributions in 1885, and he first used the phase rule in 1887 (Roozeboom 1887) to explain heterogeneous equilibria. Roozeboom subjected the data of Van't Hoff and Van Deventer to a similar analysis and found the first quintuple point: mirabilite + epsomite + bloedite + solution + gas (Roozeboom 1888*b*), located at about 22°C. Roozeboom had a clear grasp of the meaning of the phase rule (Roozeboom 1888*a*) and its implications, and he was mainly responsible for its introduction into European science and into chemistry.

> The complete heterogeneous equilibrium is expressed by a curve (p.t.), independent of the quantity of each phase present. If one follows such a curve in either direction, one arrives at a point where a new phase occurs and where a further change of temperature or pressure is not possible, unless one of the co-existing phases disappears. As soon as this happens, a new system of n + 1 phases remains, for which a curve originates in the end point of the former curve. Hence at the intersection itself n + 2 phases coexist. Since one can remove any phase one chooses without destroying the equilibrium, this point must be the beginning or end of all n + 1 curves which represent the equilibrium of each system of n + 1 phases contained within n + 2 phases (pp. 473-74).

His efforts culminated in his famous treatise on heterogeneous equilibria (Roozeboom 1901).

A teacher of Roozeboom's, J. M. Van Bemmelen, professor of chemistry at Leiden, provided the next important step in our saga. Van Bemmelen had long been interested in weathering processes, particularly the weathering of volcanic rocks to form clay minerals and laterites. In an address as rector of the University in 1889, he pointed out that studies such as those of Van't Hoff and Van Deventer could be used to elucidate the origin of the Stassfurt evaporites, a remark Van't Hoff himself recalled (Van't Hoff 1905, p. 2).

The work on bloedite (Van't Hoff & Van Deventer 1887) was followed by further studies on double salts. In 1888 W. Meyerhoffer became a student of Van't Hoff's after having worked with Ostwald in Leipzig. He investigated the copper-potassium chloride salts (Meyerhoffer 1889, 1890) and found two quintuple points. Meyerhoffer settled in Vienna (1891-96). He later was to have a decisive influence on Van't Hoff's choice of marine evaporites as a research topic and was to become his first and most permanent collaborator. During his Vienna years he wrote the first book in any language on the phase rule (Meyerhoffer 1893) and an important paper on reciprocal salt pairs (Meyerhoffer 1895). He seems to have originated the term "Phasenregel," translated with a slight but significant shift in meaning as "phase rule," and he gives clear and concrete definitions of the terms "phase" and "component" (Meyerhoffer 1890):

> *The "phase rule" and its application to the preceding solubility determinations.*
> According to T. Willard Gibbs, a heterogeneous equilibrium is completely defined when n substances which come together in an equilibrium, appear in (n + 1) phases. With respect to the term "phase," a concrete definition has been missing up to now. One could describe a phase as that part of a mixture of any bodies in any state of aggregation which can be isolated by mechanical means. . . . The equilibrium is completely defined when the compositions of the phases are independent of the amount of the individual constituents.
>
> Components ["Stoffe" in Meyerhoffer, "bodies" in Gibbs] are all those substances which come together in varying amounts and whose combinations or mutual transformations suffice to form all those compounds which play a role in the equilibrium (pp. 122–23).

These definitions are explicit and essentially those used today. They precede the full-scale introduction of Gibbs by Riecke (1890) in the following volume of the same journal. Riecke makes the first direct reference to variance: ". . . if k is the number of chemical components, then, for a given P and T, only $k + 2$ phases can coexist. If the number of phases is smaller, a corresponding number of variables remain undefined."

The power of the geometric methods developed by Gibbs was pointed out three years later by Van Rijn van Alkemade (1893), like Roozeboom, a student of the University of Leiden. With this the adoption of the phase rule and of phase theory on the continent was essentially completed.

The close connection between the work on double salts and the evolution of phase theory is further emphasized by the contribution of another student of Van Bemmelen's in Leiden, F. A. H. Schreinemakers (1891), who worked on the compound $K_2PbI_4 \cdot 2\frac{1}{2} H_2O$. Schreinemakers followed Roozeboom's analysis of bloedite. He was later to become best known for his work on ternary systems, completing the edifice left unfinished by Roozeboom (Schreinemakers 1912-24).

MARINE EVAPORITES

Meanwhile, Van't Hoff and his students continued their work on double salts. Van der Heide (1893) chose the system K_2SO_4-$MgSO_4$-H_2O. Earlier, Precht and Wittgen (1882), working at Stassfurt, had published solubility curves which exhibited unexplained kinks and inflections. Van der Heide (1893) showed that those kinks were due to the presence of the compound

$K_2Mg(SO_4)_2 \cdot 4 H_2O$, subsequently found at Stassfurt as the mineral leonite (Naupert and Wense 1893). The other double salt, $K_2Mg(SO_4)_2 \cdot 6 H_2O$, had long been known as the mineral schoenite. To determine the transition temperatures, Van der Heide used dilatometer methods as well as simple heating and cooling curves, with temperature arrests indicating phase transitions. Probably under the influence of Meyerhoffer (1893), this was the first paper from Van't Hoff's laboratory which explicitly uses Gibbs's phase rule. As was the custom since Roozeboom (1887), it is interpreted to mean that a "complete equilibrium" is established when $r = n + 1$: (r: number of phases, n: number of components). This applied, of course, to a univariant equilibrium, which was customarily examined at constant temperature.

Van't Hoff asked R. Löwenherz to continue the work of Van der Heide by adding chlorides to the system K_2SO_4-$MgSO_4$-H_2O. Löwenherz (1894) devised a projection for reciprocal salt systems—the Löwenherz projection—which was used exclusively by Van't Hoff for the subsequent Stassfurt work. It is based on plotting phase boundaries in a tetragonal prism with equilateral sides, with the square base representing the anhydrous salts and the apex H_2O. Saturated solution boundaries are then projected from the H_2O apex onto the base. In his study, based primarily on solubility determinations, Löwenherz produced the first important diagram directly applicable to the Stassfurt deposits (Fig. 1), containing information on the minerals epsomite, hexahydrate, bischofite, carnallite, sylvite, and schönite. In later years the Löwenherz projection was replaced by the Jänecke projection (Jänecke 1906, 1907), which is based on projection onto a triangle.

The next significant contribution to the subject comes from Meyerhoffer, now working in Vienna (Meyerhoffer 1895). Although concerned principally with compounds in the system Na-NH_4-NO_3-H_2O, the thrust of the investigation is now clearly spelled out: "The salt deposits of Stassfurt, Wieliczka and elsewhere, as far as they are of marine origin, cannot be interpreted in detail until the solubility and equilibrium relations of the salts of the sea have been studied systematically" (p. 854).

This passage was quoted by Van't Hoff in his first book on salt deposits (Van't Hoff 1905). Meyerhoffer (1895) goes on to say that while isothermal evaporation experiments should be used as a guide to interpretation, actual deposition may not be strictly isothermal. He also points to the possibility of solid solution in some compounds, to the modification of brine composition through the influx of other brines, and to the effects of pressure. He expresses his faith in practical terms: "The equilibrium studies are important not only for the formation of the natural salt deposits, but also for their mining. The sophisticated and economically important salt industry will gain from such studies and will change from its present empirical procedures to a period in which production will be based on purely scientific principles" (p. 855).

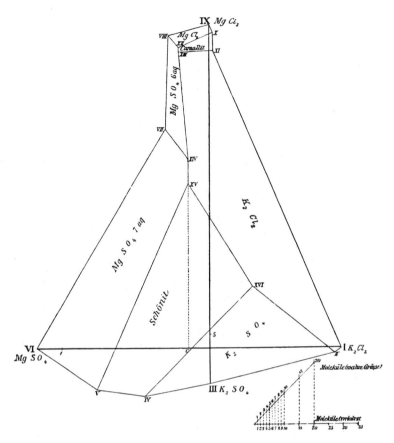

Fig. 1. Equilibrium diagram of the reciprocal system $KCl-MgCl_2-K_2$ $SO_4-MgSO_4-H_2O$. The surfaces for saturated solutions are shown in a Löwenherz projection. Copy of the original diagram given by Löwenherz (1894). Schönite is picromerite, $K_2Mg(SO_4)_2 \cdot 6 H_2O$.

MOVE TO BERLIN AND CHOICE OF RESEARCH TOPIC

Van't Hoff's new Amsterdam Institute was completed in 1891. His circle of students and collaborators continued to grow and with it his administrative duties, much to his chagrin. His restlessness became known in Holland and abroad and resulted in a call to the chair of physics at the University of Berlin (1894) as successor to Helmholtz. Planck acted as negotiator. Van't Hoff was flattered, but declined. He let it be known that a position free of administrative and teaching duties would suit him better. To underscore his dissatisfaction with his lack of freedom he suddenly resigned his chair at Amsterdam in 1895 and began a period of rest with his family in the Black Forest of

Southern Germany. As fall approached, he walked with his wife and four children across the Swiss Alps toward Lugano and the sun, using the railroad only occasionally.

Meanwhile his friends in Berlin considered an appointment to the Berlin Academy of Sciences, which would make Van't Hoff the first foreign member in over 100 years. Negotiations continued and in the winter of 1895–96 they succeeded in offering him such an appointment, coupled with an honorary professorship at the Berlin University. Van't Hoff accepted and moved to Berlin in the spring of 1896. It was not possible to suddenly put a large research institute at his disposal and the necessity to begin with modest means was crucial for the choice of a research topic. Van't Hoff himself recalls this (Van't Hoff 1905):

> When I exchanged my position in Amsterdam with that at the Berlin Academy of Sciences and the Berlin University, I gained almost unlimited time for myself and had to decide on a work plan. I wanted to continue some of my investigations of the Amsterdam period, that is oxidation mechanisms, optical activity and formation of double salts.
>
> To continue with oxidation mechanisms, delicate measurements were necessary to evaluate the electrical effects, for which my Berlin laboratory was not suitable. The splitting into optically active substances was essentially solved, as we had pursued this phenomenon in many directions. However, the formation of double salts and its application to natural salt deposits could lead to an extensive investigation (p. 1).

Meyerhoffer's direct influence on the choice is acknowledged by the quote mentioned earlier (p. 7). In fact, Meyerhoffer's persuasion was offered in person. Shortly after Van't Hoff arrived in Berlin, Meyerhoffer paid him a visit and accepted an offer to join him as collaborator. Meyerhoffer rented a house for himself in the Berlin suburb of Wilmersdorf (Uhlandstrasse 39), and this is where the laboratory was first established. A chemical analyst was hired and the investigations began with an astonishing élan. Fifty-two papers and twelve years later Van't Hoff considered the task essentially completed.

It was the first full-fledged study in experimental petrology and as such applied to a problem of the sedimentary environment. Conceived with exemplary clarity, it is still a model of how to proceed in such a case. Van't Hoff (1905) recalls the basic choice of the laboratory as follows:

> As the main thrust of the work was directed towards natural evaporite deposits, a certain restriction of the task was essential. We adopted the view that although we were confronted with a concrete problem of nature, it was desirable to maintain as broad an approach as possible. . . .
> The restriction was achieved by considering only the main constituents of the salt deposits. . . . In addition to sodium chloride, they were princi-

pally hydrated chlorides and sulfates of magnesium and potassium; they represent the main mass of the salt layers, the lower rock salt and the overlying saline formations. Next we added calcium compounds, which are important in separating the lower rock salt into anhydrite and poly-halite regions. In terms of quantity, bromine would be next, but it was eliminated as being less significant for the general problem. However, borates were considered, because they represent important occurrences as well as special problems. This gave our task a certain completeness, delin-eated as it was by the next corresponding problem of nature: the forma-tion of silicates (pp. 2–3).

The last remark points directly at the Geophysical Laboratory of the Carnegie Institution, which was established about the time Van't Hoff wrote his summary and which will be discussed later. The credo of the modern experimental petrologist is to be found in the same introduction (Van't Hoff 1905): "The answers to these questions are pursued as far as their solutions are contained within the tasks formulated in the broadest sense. However, in the specific execution, contact is maintained with the natural environment in order to avoid unnecessary detailed work" (p. 3). Van't Hoff practiced this conviction by keeping in close contact with the natural deposits, using Kubierschky and Precht as geologic guides.

VAN'T HOFF'S CONTRIBUTION TO MARINE EVAPORITES

The work commenced in 1896 and was nearly completed in the spring of 1908. The approach, in general and in detail, remained consistent throughout and essentially followed the procedures established in the first paper on double salts (Van't Hoff & Van Deventer 1887). Van't Hoff and Meyerhoffer first turned their attention to the systems $MgCl_2-H_2O$, $KCl-MgCl_2-H_2O$ and $CaCl_2-MgCl_2-H_2O$ and the minerals bischofite, carnallite, and tachhydrite. Solution compositions and phase transitions were determined over a wide temperature range (below 0 to 200°C), the former by chemical analysis and the latter by simple heating and cooling curves. The tachhydrite paper (IV)[2] contains the first direct geologic application. Below 21.95°C tachhydrite was found to decompose to bischofite and calcium chloride:

$$2MgCl_2 \cdot CaCl_2 \cdot 12\,H_2O \rightleftharpoons 2MgCl_2 \cdot 6H_2O + CaCl_2 \cdot 6H_2O + 6H_2O$$

tachhydrite   bischofite

Tachhydrite is found within anhydrite in the carnallite zone. Van't Hoff and Meyerhoffer (IV) remark: "With respect to the formation of natural salt

[2] Roman numerals refer to the numbering system used in Van't Hoff (1912) for the contributions which appeared in the *Annals of the Prussian Academy of Sciences*.

deposits it is especially important to realize that the temperature 21.95 represents a lower limit, and that the occurrence of tachhydrite indicates a temperature above this limit" (p. 40).

After this preliminary work, Van't Hoff addresses himself to the important reciprocal system $KCl-MgCl_2-K_2SO_4-MgSO_4-H_2O$ previously studied by Löwenherz (1894). He now restricts himself to 25°C and to a detailed review of crystallization paths. Löwenherz (1897) had pointed out some large discrepancies in his own determinations. Van't Hoff carefully repeated the experiments and found that they were due to rimming of sylvite by carnallite and hence there was lack of equilibrium between sylvite and the solution (p. 55). Water vapor pressures of the saturated solutions are reported next in contribution VI, which also contains the remark: "Finally, the vapor pressures . . . may be connected with the atmospheric humidity existing during the formation of the evaporites" (p. 57). The next five papers (VII-XI) are concerned with the Löwenherz system saturated with respect to NaCl at 25°C. They deal with the minerals halite, sylvite, carnallite, bischofite, epsomite, picromerite, aphthitalite, thenardite, and bloedite and lead up to a review of the evaporation of sea water at 25°C. The sequence of precipitates found was halite → epsomite → hexahydrate → sylvite → carnallite → bischofite. The absence of kieserite, loeweite, kainite, and langbeinite was blamed on the lack of temperature variation. These conclusions were to be modified frequently during the progress of the work.

The first check was made through the vapor pressures of the saturated solutions (XIX) at 25°C. This pointed to the stable existence of kainite at 25°C (XXI) and the replacement of sylvite by kainite in the evaporation sequence.

Meanwhile, Van't Hoff decided to look at the calcium-bearing phases as well, such as gypsum (XVIII), anhydrite (XXIV), glauberite (XV, XLII), syngenite (XX), and polyhalite. The presence of these relatively insoluble compounds did not affect the compositions of the saturated solutions, but it was important to determine which of the phases was stable for a specific set of conditions. The glauberite field was not intersected by evaporating sea water at 25°C, but that of gypsum was. The phase relations between gypsum, bassanite, and anhydrite occupied Van't Hoff and his collaborators for a number of years (XVIII, XXII, XXIV, XXXIII, Van't Hoff et al. 1903). Hardie (1967) has reviewed their results for the gypsum-anhydrite transition and compared them with subsequent determinations (pp. 187-88). It is indicative of Van't Hoff's brilliance as an experimenter that his values for the transition temperature are closest to the presently accepted set of values (Hardie 1967, Fig. 5). The approach used by Van't Hoff is clearly outlined in the first paper (XVII) concerned with gypsum and bassanite, where he demonstrates the power of thermodynamic reasoning for the experimentalist.

He begins (p. 135) with the Van't Hoff equation:[3]

$$\text{``}\frac{d1p}{dT} = \frac{q}{2T^2}\text{,''}$$

whereby, if we deal with water:

(1) $\dfrac{d1p_w}{dT} = \dfrac{q_w}{2T^2}$ where q_w is the latent heat of vaporization for 18 kg of water.

Using this equation for gypsum:

(2) $\dfrac{d1p_g}{dT} = \dfrac{q_g}{2T^2}$ where q_g corresponds to the latent heat of vaporization for 18 kg water bound in gypsum.

By subtracting (1) from (2):

$$\frac{d1\dfrac{p_g}{p_w}}{dT} = \frac{q_g - q_w}{2T^2} = \frac{q}{2T^2}$$ where q corresponds to the heat evolved

when 18 kg water combine with bassanite to form gypsum. If, in first approximation, the value of q is assumed to be constant, we obtain by integration

$$1\frac{p_g}{p_w} = -\frac{q}{2T} + \text{const}$$

and this formula allows us to extend our determinations to other temperatures. If we use as a basis for our calculations

$T = 273 + 101.45°$ $p_g = 758.8$ $p_w = 800.4$
$T = 273 + 25$ $p_g =$ 9.1 $p_w =$ 23.52 .

We obtain $q = 2614$ and

$$\log p_g = \log p_w + 1.493 - \frac{567.7}{t+273}$$ for the pressure of water bound in gypsum at temperature t."

[3] In the modern form $\dfrac{d\ln P}{dT} = \dfrac{\Delta H}{RT^2}$.

This corresponds to the current formalism

$$\log a_{H_2O} = \log \frac{f_{H_2O}}{f^o_{H_2O}} = A - \frac{B}{T}$$

where a_{H_2O} and f_{H_2O} are the activity and fugacity respectively of H_2O in solution, and $f^o_{H_2O}$ is the fugacity of pure water at the same P and T.

The same approach was used to fix the gypsum-anhydrite transition at 1 atm pressure. This was accomplished by measuring the vapor pressure of the saturated solution at two points, one saturated with halite and the other saturated with $NaBrO_3$, which yielded the relationship

$$\log P^{sol.}_{H_2O} = \log P^{water}_{H_2O} + 1.486 - \frac{500}{T^\circ K}$$

and a transition temperature of $63.5^\circ C$ in calcium sulfate solutions. This compares very well with the modern value of $58^\circ C$ (Hardie 1967).

Next Van't Hoff returned to the more soluble salts, but now to those phases which were not found at $25^\circ C$. Langbeinite was encountered at 37° (XXV), loeweite at 43° (XXVI), vanthoffite at 46° (XXXI), and at 83° kainite decomposed to a mixture of kieserite + sylvite (XXXII). In this manner, Van't Hoff constructed a table of metamorphic changes for evaporite minerals (XXXII) (p. 327), based on a very accurate temperature calibration (see Fig. 2). Apparently, Van't Hoff decided that $83^\circ C$ was a natural upper limit for his investigations, and he next determined vapor pressures and compositions of the saturated solutions at that temperature (XXXIV and XXXV). With this framework erected, it was possible to discuss mineral assemblages as they change from 25° to $83^\circ C$ (XXXVI). Van't Hoff distinguished three intervals: $25-37^\circ$, $37-55^\circ$, and $55-83^\circ C$. The first interval is characterized by the disappearance of picromerite, epsomite, and hexahydrate; the second by the appearance of langbeinite, loeweite, and vanthoffite; and the third by the disappearance of bloedite, leonite, and kainite.

The crucial test for experimentalists, a comparison of all stable experimental assemblages with naturally occuring assemblages, was not possible until a complete list of natural assemblages became available. When the latter was published (Erdmann 1907), Van't Hoff (1909) immediately drew up a comparison. He scored very well, as of the 230 or so possible assemblages only 14 were unexpected, and most of these could be accounted for, while 66 of the expected 103 had actually been described. This result must be judged a major triumph for Van't Hoff's approach, his care, and his tenacity. It certainly has not been surpassed since.

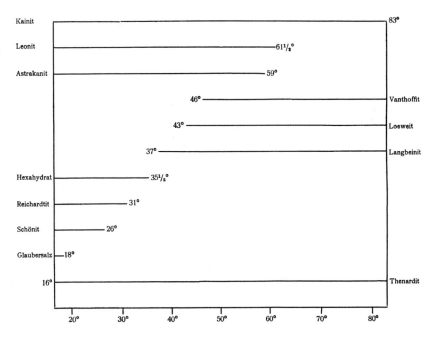

Fig. 2. Temperature grid for common marine evaporite minerals. Copy of original diagram given by Van't Hoff (1912).

Van't Hoff extended his studies to include some equilibria below 25°C (XXXIX) and then embarked upon an investigation of borate minerals, which occupied him during most of the remaining years. We need not review these here. They followed by now well-established paths and they are less significant for the interpretation of the marine evaporites.

Van't Hoff summarized his contributions in two small volumes (1905, 1909), the first dealing with the more soluble salts, and the second with the calcium-bearing phases and the borate minerals. The material is organized differently, with an attempt at greater generality. The first volume (1905) contains a section on secondary changes, in which the effects of adding water are considered; and a section on temperature determinations based on mineral assemblages. The second volume (1909) extends these applications and also lists remaining tasks. With respect to varves, Van't Hoff says: "The fact that anhydrite and polyhalite alternate with halite in so-called varves could be connected with the opposite effect of temperature on their solubility, which for halite increases somewhat with temperature and for anhydrite decreases" (p. 41).

The correspondence between experiments and nature is characterized as follows:

These results correlate with natural relationships qualitatively, but not at all quantitatively, which indicates that salt deposits are not formed by simple evaporation of sea water (p. 42).

The discrepancy is two-fold, because the original amount of anhydrite in nature is much larger than the overlying halite and because the latter contains considerably more anhydrite than it should. If one assumes that during evaporation sea water is added, the relationship between total amount of anhydrite and halite is not changed. . . . The thickness of the lower gypsum and anhydrite beds compared with halite requires an additional assumption: either that not only sea water was added, but brine was lost, or that gypsum-bearing river water was added, as Winther has suggested (p. 77).

In the summary, Van't Hoff states:

In concluding these investigations I am fully aware of having made but a step, and at the outset I had not intended to do more, as we were dealing essentially with a chemical point of view. I restricted myself deliberately, but always kept in touch with natural occurrences, either through personal contacts, for which I am indebted to Misters Kubierschky and Precht, or through repeated visits to the mines. . . . However, in order to achieve the desirable integration with the natural occurrences, an extensive study is still needed to bridge chemistry and geology. A significant contribution in this direction could be made by the association for the study of natural salt deposits, which was formed at the suggestion of Mr. Rinne (p. 75).

VAN'T HOFF AND HIS DISCIPLES

The last remark points directly to the fate of Van't Hoff's investigations in Germany. The "association for the study of natural salt deposits" was formed in 1906 with the express aim to continue the work and to broaden its base. Van't Hoff himself considered his task nearly completed and was glad that its continuation was assured. He included an account of the history of the association and its first two annual reports in his last, the 52nd, paper (Van't Hoff 1912, LII). The completion was timely, as Meyerhoffer, his chief collaborator, had died in 1906 and Van't Hoff himself was ordered to cease all experimental work in the spring of 1908 for medical reasons. There was still time to finish the summary (Van't Hoff 1909) and for a move to the suburbs where he hoped to continue his work on fermentation. But illness hampered his progress and he died March 1, 1911.

In contemplating Van't Hoff's contribution as a whole, one must be impressed by the resourcefulness in experimental techniques, by the almost dogged persistence with which the originally conceived project is pursued, and by the virtual absence of new theory. It is remarkable that Van't Hoff, though he was familiar with the phase rule, even lectured on it (Van't Hoff

1902), did not use it in a single instance throughout his study of salt deposits. His attitude is best summarized by this quote: "It is a pity, that gradually a certain exaggeration has developed with respect to the consequences of the phase rule, regardless of the importance of its content. . . . The large significance of the phase rule lies not so much in its value as a guide during the investigation, but rather as a pedagogical tool for the treatment and classification of chemical equilibria" (pp. 4252-53).

This extremely cautious stand is perhaps understandable, if we remember that Van't Hoff in his "law of the incompatibility of condensed systems" had been concerned with phase-rule problems at a time when he was still ignorant of the phase rule. Also, his work on optically active substances dealt with cases which are particularly difficult to handle with the standard phase-rule application. It is a pity that he did not trust Gibbs's thermodynamics more and that he did not use the geometric methods based on Gibbs's free energy surfaces, which were introduced on the continent by Van Rijn van Alkemade (1893) in his own journal. Van't Hoff's influence was so strong, that this attitude persisted for over half a century in evaporite studies. As late as 1915 one of his students, A. E. Boeke, in a much admired text on chemical petrology (Boeke 1915) says: "However we should not forget that the representation of the relationships by zeta surfaces is only an image of experience, which has been derived from the thermodynamics of isotropic bodies (gases and liquids) and extended, perhaps not with complete justification, to the crystalline state" (p. 101). The question of why Van't Hoff, who became famous by introducing theory into the largely empirical field of chemistry, abandoned theoretical work altogether is difficult to answer. Fischer (in Van't Hoff 1912) mentions that many of Van't Hoff's friends interpreted the Berlin period as a time of rest, and perhaps even of exhaustion in Van't Hoff's life. In contrast, Fischer suggests that Van't Hoff felt he had completed the theoretical framework and that it was now necessary for him to return to observation and experiment before further generalizations were possible. I would prefer a slightly different interpretation. The effort was so massive and uncompromising that there could have been little thought of rest for Van't Hoff. It is true, however, that he concerned himself very little with theory, nor, for that matter, with any other problem outside the chosen task. This is very unusual for a man with the breadth of interests Van't Hoff had demonstrated earlier. Could it be that he had committed himself to a problem which was very much larger than anticipated, and, once committed, he felt obliged to complete it? Van't Hoff himself gives some hints in this direction (Jorissen & Reicher 1912, pp. 73-74): "For me the work we began once again was only in part connected with salt deposits, but the increasing fascination and the magnitude of the problems slowly displaced other thoughts. . . . At first everything went smoothly beyond expectation. . . . But our hopes were not fulfilled and a visit to the Neu-Stassfurt salt mines with Mr. Precht made it clear, that at the moment we had more to learn from Geology and Mineralogy than vice versa." Even at the end, he showed little evidence of mental exhaustion and lost originality as he tackled an entirely different and equally taxing new problem, that of fermentation, and he was stopped only by failing health.

Van't Hoff's disciples continued to work entirely in his spirit, foremost among them d'Ans, Boeke, Jänecke, and Rinne. They refined the edifice and continued to keep Germany in the forefront of evaporite research. A certain narrowing of outlook was inevitable. This is clearly expressed in the introduction to Jänecke's book on salt deposits (Jänecke 1915): "While in a book on the origin of coal deposits the geologic parts of their formation must be stressed more than chemical questions, the opposite is true for potash deposits. Here the physical-chemical aspects of precipitation and transformation are much more important for their genesis than geologic considerations. . . . furthermore, with respect to the purely geologic processes there can hardly remain any differences of opinion" (p. VI).

In other words, the problem could now be reduced to a purely chemical question and contact with natural deposits was no longer essential. Fortunately, this attitude was not universal and the books by Lotze (1957), Borchert (1959), and Braitsch (1962) preserve a careful balance between laboratory data and geologic arguments. However, even the most recent experimental studies, such as those of Autenrieth (1958), Baar and Kühn (1962), Braitsch and Herrmann (1962) clearly show their debt to Van't Hoff. The first radical departure was contributed by another physical chemist, Sillen (1959), when he applied the phase rule to problems of the composition of sea water.

The impact of Van't Hoff's work on other branches of geology is fairly easy to assess. He did not influence the study of sedimentary rocks in general, which remained descriptive and largely antiphysical chemistry. This attitude remained unchallenged until the paper of Krumbein and Garrels (1952). Van't Hoff's true heirs are Day and Allen (1905), Goldschmidt (1911), and Bowen (1913) and their studies of igneous and metamorphic rocks. As late as 1911, Rinne (1911) complains that Van't Hoff's monumental contribution remains unappreciated. After reiterating the fact that mineral formations from aqueous solutions and from silicate magmas follow the same laws, a viewpoint arrived at half a century earlier by Scheerer, Sorby, and Bunsen, he says:

A survey of the petrologic literature shows that this thought of Bunsen . . . has been accepted only very rarely by petrologists. The fate of the physical-chemical studies of the Zechstein salts by Van't Hoff demonstrates this very clearly. In this study . . . petrology possesses the first example of a fundamental, systematic and experimentally exact study and insight into the conditions of formation of a large rock unit, arrived at through precise physical-chemical reasoning. . . . Nevertheless, this monumental contribution has hardly been noticed by the leading petrologists (p. 183).

VAN'T HOFF AND IGNEOUS PETROLOGY

Meanwhile, partly inspired by Van't Hoff's success, a parallel development was taking place which was to change the relationship between petrology and chemistry for good: the founding of the Geophysical Laboratory of the Carnegie Institution of Washington. Let us turn back to the beginning of the century and consider this event more carefully.

In June of 1900, C. F. Becker of the U.S. Geological Survey was put in charge of the Division of Physical and Chemical Research. The Division consisted of two laboratories, one on chemistry under F. W. Clarke and one on physics as yet without a leader. Becker was not new to geophysical research within the Survey, as he stated himself (Becker 1905):

It was in recognition of the need for researches in physics which would throw light on geological problems that Dr. Carl Barus was appointed physicist on my staff in the United States Geological Survey as far back as 1880, and that a physical laboratory was established under that Survey in 1882. This was discontinued in 1892, not because its importance was underestimated by the Director, but on account of a failure of appropriations. The laboratory was re-established in 1901 because it was felt that without the aid to be derived from physical determinations the efficiency of the Survey must suffer (p. 6).

In October 1900, A. L. Day, then assistant at the Physikalisch-Technische Reichsanstalt, Charlottenburg, Berlin, was given a three-month temporary appointment as physical geologist, and in January 1901 a permanent appointment, to build a new physical laboratory in Becker's Division. C. D. Walcott was director of the survey at that time (Walcott 1901). Day had studied physics at Yale and Berlin and had spent the last three years at the Reichsanstalt, working with Holborn on high-temperature gas thermometry (Holborn & Day 1899, 1900). Day spent the first year in Washington setting up equipment in part borrowed from the Smithsonian Institution and in part made to his specifications in Charlottenburg. The first geologic problem presented to him by a Survey member was concerned with the ribbon structure of auriferous ores. This work eventually led to a publication on the force of crystallization (Day & Becker 1916).

"Dr. A. L. Day and Mr. Van Orstrand were employed in perfecting equipment of the physical laboratory, in numerous experiments on the linear force exerted by growing crystals, and on the elastic properties of solids dealt with in a novel manner" (Walcott 1902, p. 121).

Day continued his interest in high-temperature work by building a furnace:

> ". . . and the furnace is now in constant use for all temperatures up to 1600°C. The method enables both the temperature and the conductivity of the material under investigation to be simultaneously measured . . . and leaves little to be desired as a furnace for mineral investigation. As soon as the furnace had been brought into good working order, the melting temperatures of the principal rock-forming minerals were taken up, beginning with representative feldspars" (Walcott 1903, p. 127).

Meanwhile Van Orstrand had left Day to work with Becker and E. T. Allen had joined him. Next year Walcott (1904) reports: "A. L. Day and E. T. Allen completed the study of the thermal properties of the plagioclase feldspars at ordinary pressure" (p. 100); and "The Carnegie Institution made a grant of money to Messrs. Becker and Day for researches respectively in elasticity and high-temperature work, laboratory space to be furnished by the Survey" (p. 103).

With this initial support of the Survey's efforts by the Carnegie Institution the fate of experimental petrology was profoundly affected. The Carnegie Institution of Washington, founded in 1902, had been casting around for ways of supporting research in geophysics. C. D. Walcott was its secretary, and he was also a member of its committee on geophysics, together with Barus, Chamberlin, Michelson, Van Hise, and Woodward. This committee submitted a comprehensive report on the need for a laboratory of geophysics in Washington (Gilman 1903). In their justification they leaned heavily on the success of Van't Hoff and Ostwald:

> Until recently the natural sciences and physical sciences have been handled as if almost independent of each other. The ground between has been largely neglected. The occupancy of this ground is certain to lead to important results. The order of results to be expected is illustrated by the great advances which have recently come from occupying the middle ground between astronomy and physics, and between physics and chemistry. . . . Chemistry and physics for a long time were pursued as independent sciences. The rapid rise of physical chemistry has shown how wonderfully fruitful is the ground between the two" (p. 27).

Becker, in an appendix to this report, mentions: "That magmas are solutions is known, but scarcely anything is known about them. Thus, vulcanism implies researches on the nature of igneous solutions. . . . A special brand of this subject, but most important, is the study of aquo-igneous fusion and the solutions resulting from it" (pp. 46–47).

Walcott, as the secretary of the Institution, had asked world-reknowned scientists to comment on the proposal. Answers included in the report were from Poincaré, Kelvin, Suess, Mach, Becke, Kohlrausch, Van't Hoff, G. H. Darwin, and Nernst. Van't Hoff himself says:

... an investigation in that direction might prove of the highest value, if made in a systematic way and continued for some years.

The special problem, which I mean, deserves attention, is the physical chemistry of high temperatures applied to the chief constituents of the earth's crust, silicates in the first instance. To express myself more clearly, I add that two great problems concerning geophysics may in the present state of our knowledge be solved, viz., the evaporation of complex solutions, which have produced systemic deposits, such as salt layers, etc.; and secondly, the cooling down of molten masses, that have produced the volcanic and plutonic formations.

With the first problem, by far the easier one as regards apparatus, etc., I have been occupied for more than six years, and a series of twenty-six publications in the *Annals of the Prussian Academy of Sciences* (1897–1902) shows how far these researches have been carried out (p. 65).

In March 1903, Walcott requested Van Hise and Becker to assemble further supporting material for the justification and construction of the proposed geophysical laboratory. Van Hise reported (Van Hise 1903): "Van't Hoff, Ostwald, and others, seeing that there was a great unoccupied field between physics and chemistry, began the occupation of it. The great results reached by these men placed their names very high in the roll of those who have contributed fundamental ideas to science" (p. 174).

And he reiterated his conviction: "The purpose of a geophysical laboratory is to take possession of the vacant ground between geology and physics and geology and chemistry. So long as geology remained a descriptive science, it had little need of chemistry and physics; but the time has now come when geologists are not satisfied with mere descriptions. They desire to interpret the phenomena they see in reference to their causes—in other words, under the principles of physics and chemistry" (p. 174).

Again using Van't Hoff as the shining example, he says: "Another class of investigations is the artificial production of minerals and rocks from aqueous solutions. This involves a study of natural solutions, both those of the sea and those in openings of rocks, in order to determine the conditions under which minerals crystallize from such solutions. Already the study of natural solutions with reference to the crystallization of salt and gypsum has been undertaken by Van't Hoff. This great chemist has reached many important results, but he points out that very much remains to be done" (pp. 179–80).

This seems to have been sufficient justification for the trustees of the Carnegie Institution. As a first step they made direct grants to Becker and

Day of the U.S. Geological Survey. Day's grant of $12,500 was "to increase and extend the work of the high-temperature research in certain particular directions: (1) by increasing the scope of the researches of the rock-forming minerals at extreme temperatures; (2) by providing for experimentation at extreme pressures as well; and thereby (3) to develop apparatus for experiments upon aqueo-igneous fusion" (Gilman, 1904, p. 80).

The grant to Day was renewed the next year at $15,000—and Day reports (1905): "The investigation of the lime-soda feldspar group which was begun in the Geological Survey has been finished, after nearly three years' work upon it. The investigation has shown that the lime-soda feldspars form a continuous series of mixed crystals capable of stable existence in any proportion of the two component minerals" (pp. 225–26).

The results were published as Publication No. 31 of the Carnegie Institution of Washington (Day & Allen 1905) with an introduction by G. F. Becker, and optical studies by J. P. Iddings. This great contribution to experimental petrology is both a worthy successor to Van't Hoff's work and a counterpart to it in the high-temperature field. Its impact, however, was much more immediate and extensive, probably because it addressed itself to a different audience within geology, one willing and eager for this new language in igneous petrology.

The Carnegie Institution trustees also expressed their confidence in Day's work, both by renewing his grant at $17,500 and by appropriating, in December 1905, $150,000 for a geophysical laboratory with Day as its director. Of his last year's work at the U.S. Geological Survey, Day (1906) reports: "With the work of the present year our studies of mineral fusion and solution in the laboratory may be said to have passed beyond the preliminary stage. It has been found thoroughly practicable to study several of the important problems in mineral formation by applying the principles and methods of physics and physical chemistry at the temperature where the formation actually occurs, and to carry out the quantitative determinations with an accuracy entirely comparable with the more conventional physical and chemical research at ordinary temperatures" (p. 177). He lists the twelve papers published during the past year by members of his laboratory, which included Allen, Sheperd, White, and Wright, all subsequently members of the Geophysical Laboratory's original staff. The laboratory was constructed during 1906–07 and occupied by the Geological Survey group, equipment, and personnel, during the last few days of June 1907. With its opening, the future of experimental petrology was assured, if not for the field of sedimentary processes, for which it originally was established by Van't Hoff. Igneous rocks were its primary concern at first, and in N. L. Bowen it could boast of the most famous geologic experimentalist. Bowen built directly upon the work of Day and Allen (Bowen 1913) and became, through his mastery of both classical petrology and modern experimental approaches, its most effective proponent.

The founding of the Geophysical Laboratory was, of course, not a direct result of Van't Hoff's efforts on behalf of evaporites, but its indebtedness to his general approach and his success both in physical chemistry and experimental petrology is obvious in the statements of Van Hise and the geophysics committee.

Day himself does not directly pay tribute to Van't Hoff, but Van't Hoff (in Jorissen & Reicher 1912, p. 76) refers to a thank-you note from Day which he received in 1906 and in which Day told him of the founding of the Geophysical Laboratory. To Van't Hoff this institution, devoted to the study of silicates, was analogous to the "Kaliverband" (association for the study of natural salt deposits), which was to continue where he left off with the study of evaporite deposits.

VAN'T HOFF AND METAMORPHIC ROCKS

One further influence of Van't Hoff on geology remains to be traced: that on the study of metamorphic rocks. In his classic memoir on the contact metamorphism of the Kristiania region (Goldschmidt 1911), Goldschmidt clearly formulates his debt to Van't Hoff:

> The fundamental law which governs the relationship between chemical composition and mineral assemblage of contact rocks is the *phase rule.* If we have as starting materials a certain number of compounds for instance those of a shale, then we can calculate the number of minerals to be found in the products of contact metamorphism. However, in its most general formulation, the phase rule of Willard Gibbs[1] cannot be applied with ease and confidence, if we are dealing with a system of many different substances. For our needs, in considering hornfelses derived from shales, the following substances are important: silica, alumina, magnesia, and lime. Systems of four components present severe difficulties, if treated with the most generalized phase rule. I prefer, for reasons of simplicity, to apply the phase rule in a special form, namely in the form of the laws of the formation of double salts.[2] These laws are contained within the phase rule, but they are especially suited to interpret the reactions of contact rocks. *We consider that part of a rock which is in a reactive state as a saturated solution, that is a solution saturated with respect to the relevant contact minerals*; the minerals produced by the contact metamorphism we consider as *precipitates* of that solution.

[1]From the point of view of the mineralogist, Willard Gibbs's phase rule seems to me to have the following formulation:

The maximum number n of solid minerals which can coexist simultaneously and stably, is equal to the number n of individual components contained in these minerals (if one disregards the singular temperatures of transition points).

[2]Van't Hoff's laws of double salts can be found in the following publications: *Studies in chemical dynamics*, Leipzig, 1896; *Lectures on formation and decomposition of double salts*, Leipzig, 1897; Lectures in theoretical and physical chemistry, Braunschweig, 1898 (pp. 123–24).

This stunning quote illustrates how difficult it was even for the most enlightened geologists to cope with the phase rule, perhaps because it was formulated in such abstract and mathematical language. Goldschmidt obviously preferred the more tangible result of Van't Hoff and he proceeded to analyze the mineral assemblages of his hornfelses in the light of the "law of the compatibility of condensed systems." His approach was so obviously successful that it became a landmark in the study of metamorphic rocks and an important step toward Eskola's eventual formulation of the facies principle (Eskola 1920).

CONCLUSIONS

In tracing the evolution of Van't Hoff's ideas, we learned that after an early interest in spatial considerations he turned to the study of chemical kinetics and chemical equilibria. He was struck by the differences between physical and chemical equilibria. Phase transformations belonged to the latter category, and for these he formulated his "law of the incompatibility of condensed systems." In taking the opposite road, Gibbs had earlier considered all equilibria to be subject to the same strictures and all phases to be equivalent, regardless of their state of aggregation. This difference explains the greater power and generality of the phase rule.

Van't Hoff extended his law to double salts and to test the propriety of this extension he chose bloedite, a constituent of the Zechstein evaporites. The bloedite study contains many of the ingredients of the later work. The significance of his contribution for the understanding of the Stassfurt deposits was pointed out to Van't Hoff by Van Bemmelen. Van't Hoff's students, Meyerhoffer, Van der Heide, and Löwenherz continued the investigations, and Meyerhoffer in particular was aware of the geologic implications. Meyerhoffer joined Van't Hoff in Berlin when the latter accepted an appointment to the Berlin Academy of Sciences.

Van't Hoff and his collaborators labored for the next twelve years exclusively on the elucidation of the mineral equilibria which govern the Stassfurt evaporites. The study is characterized by experimental elegance, clearly formulated goals, and enviable persistence.

It is the first systematic contribution to experimental petrology. At all times, the problem was perceived as geologic in nature and the laboratory results were checked against natural assemblages whenever possible. The phase rule was not used, nor, for that matter, was chemical thermodynamics, except for Van't Hoff's equation. However, the work of Van't Hoff and Deventer was indirectly involved in the evolution of phase theory by Roozeboom, Van Rijn van Alkemade, and Schreinemakers. Meyerhoffer himself wrote the first text explicitly devoted to the phase rule.

The impact of Van't Hoff's study was enormous, but it was restricted to those geologists willing and able to cope with chemistry. Foremost among them were igneous petrologists who had long since accepted chemical arguments for classification purposes. I consider the Geophysical Laboratory program to be the most direct heir of the Van't Hoff approach. Although the shape of that program was formulated independently by Van Hise, Becker, Day, and others, the inspiration they derived from Van't Hoff's successes is clearly acknowledged. The study of the fusion of plagioclases by Day and Allen, which directly lead to the authorization for the Geophysical Laboratory, is the igneous counterpart to Van't Hoff's low-temperature experimental petrology. On metamorphic petrology, too, Van't Hoff left his mark, with V. M. Goldschmidt acting as his disciple. The interpretation of the Kristania contact rocks was explicitly based on Van't Hoff's double-salt law in preference to the phase rule.

Sedimentologists remained unaffected and continued their preoccupation with description and classification. Chemical arguments remained subordinate in their work and of an elementary nature, underscoring the chasm between "hard" rocks and "soft" rocks. This gulf is only closing now through the blossoming of experimental petrology and geochemistry since World War II. At last the generality of the point of view of Gibbs is being accepted. If Van't Hoff's contribution had been appreciated fully at the time, this could have happened seventy years earlier.

OUTLOOK

Where do we go from here in marine evaporite research? Obviously Van't Hoff's success has remained a model for many of us and it has dominated our thinking to the present day. Marine evaporites are ideally suited to the chemical approach, because their minerals can readily be precipitated from solutions. And yet even the most recent studies have not settled some basic geologic controversies, such as the question of modern analogues or the arguments of deep- versus shallow-water deposits. To settle such questions we need a broader base, one which includes sedimentary processes and the depositional environment as well as a concern with chemical equilibria. This reorientation does not diminish the significance of Van't Hoff's contribution in any way, but it simply acknowledges the fact that we may have followed in his footsteps for too long. Van't Hoff himself would surely agree with this assessment and would welcome such a change in direction.

REFERENCES

Arrhenius, S. 1887. *Über die Dissociation der in Wasser gelösten Stoffe. Zeitschr. phys. Chem. 1*, 630–48.
Autenrieth, H. 1958. Untersuchungen am sechs-Komponenten-System K, Na, Mg, Ca, SO_4, Cl, H_2O mit Schlussfolgerungen für die Verarbeitung der Kalisalze. *Kali und Steinsalz 2*, 181–200.

Baar, A., & Kühn, R. 1962. Der Werdegang der Kalisalzlagerstätten am Oberrhein. *Neues. Jahrb. Mineral. 97*, 289-336.

Becker, G. F. 1905. Introduction to: The Isomorphism and thermal properties of the feldspars. *Carnegie Inst. Washington Publ. No. 31*, 5-12.

Boeke, H. E. 1915. *Grundlagen der physikalisch-chemischen Petrographie.* Gebr. Berlin: Borntraeger, 428 pp.

Borchert, H. 1959. *Ozeane Salzlagerstätten.* Berlin: Gebr. Bornträger, 237 pp.

Bowen, N. L. 1913. The melting phenomena of the plagioclase feldspars. *Am. Jour. Sci. 35*, 577-99.

Braitsch, O. 1962. *Entstehung und Stoffbestand der Salzlagerstätten.* Berlin: Springer-Verlag, 232 pp.

Braitsch, O., & Hermann, A. G. 1963. Zur Geochemie des Broms in salinaren Sedimenten, Teil II: Die Bildungstemperaturen primärer Sylvin-und Carnallit-Gesteine. *Geochim. Cosmochim. Acta 28*, 1081-1109.

Day, A. L. 1905. Mineral solution and fusion under high temperatures and pressures. *Carnegie Inst. Washington Yearbook 4*, 224-30.

_____. 1906. Mineral solution and fusion under high temperatures and pressures. *Carnegie Inst. Washington Yearbook 5*, 177-79.

Day, A. L., & Allen, E. T. 1905. The isomorphism and thermal properties of the feldspars, Part I. *Carnegie Inst. Washington Publ. No. 31*, 13-75.

Day, A. L., & Becker, G. F. 1916. Die lineare Kraft wachsender Kristalle. *Centralblatt Mineral. 23*, 337-46.

Erdmann, E. 1907. Chemie und Industrie der Kalisalze, *in: Festschrift zum X. Allgemeinen Deutschen Bergmannstag zu Eisenach. Landesanstalt: Verl. Königl. Geol.* pp. 1-123.

Eskola, P. 1915. On the relation between chemical and mineralogical composition in the metamorphic rocks of the Orijärvi region. *Comm. géol. Finlande Bull. 44*, 1-145.

_____. 1920. The mineral facies of rocks. *Norsk geol. tidsskr. 6*, 143-94.

Fischer, E. 1912. Gedächtnisrede auf Jacobus Henricus Van't Hoff, *in:* J. H. Van't Hoff, *Untersuchungen über die Bildungsverhältnisse der ozeanischen Salzablagerungen, insbesondere des Stassfurter Salzlagers.* Leipzig: Akad. Verlagsges., 374 pp.

Gilman, D. C. 1902. Report of Advisory Committee on Geophysics. *Carnegie Inst. Washington Yearbook 1*, 26-70.

_____. 1904. Geophysical Research. *Carnegie Inst. Washington Yearbook 3*, 80-82.

Goldschmidt, V. M. 1911. Die Kontaktmetamorphose im Kristiniagebiet. *Kristiania Vidensk. Skr., Math.-Naturv. Kl.1*, 1-483.

Hardie, L. A. 1967. The gypsum-anhydrite equilibrium at one atmosphere pressure. *Am. Mineral. 52*, 171-200.

Holborn, L., & Day, A. L. 1900. On the gas-thermometer at high temperatures. *Am. Jour. Sci. 10*, 171-200.

Jänecke, E. 1906. Uber eine neue Darstellungsform der wässerigen Lösungen zweier und dreier gleichioniger Salze, reziproker Salzpaare und der Van't Hoffschen Untersuchungen *über ozeanische Salzablagerungen. Zeitschr. Anorg. Chem. 51*, 132-57.

_____. 1907. Über eine Darstellungsform der Van't Hoffschen Untersuchungen über ozeanische Salzablagerungen II. *Zeitschr. anorg. Chemie. 52*, 358–67.

_____. 1915. *Die Entstehung der Deutschen Kalisalzlager.* Braunschweig: Vieweg und Sohn, 111 pp.

Jorissen, W. P., & Reicher, L. T. 1912. *J. H. Van't Hoff's Amsterdamer Periode.* Helder: C. De Boer, 106 pp.

Krumbein, W. C., & Garrels, R. M. 1952. Origin and classification of chemical sediments in terms of pH and oxidation-reduction potentials. *Jour. Geology 60*, 1–33.

Lotze, F. 1957. *Steinsalz und Kalisalze.* I. Teil. Berlin: Gebr. Borntraeger, 936 pp.

Löwenherz, R. 1894. Über gesättigte Lösungen von Magnesiumchlorid und Kaliumsulfat oder von Magnesiumsulfat und Kaliumchlorid. *Zeitschr. phys. Chemie 13*, 459–91.

_____. 1897. Über gesättigte Lösungen von Magnesiumchlorid und Kaliumsulfat oder von Magnesiumsulfat und Kaliumchlorid. *Zeitschr. phys. Chemie 23*, 95–96.

Meyerhoffer, W. 1889. Über die reversible Umwandlung des Cupribikaliumchlorids. *Zeitschr. phys. Chemie 3*, 336–46.

_____. 1890. Über die gesättigten Lösungen der Verbindungen von Cuprichlorid mit Kaliumchlorid. *Zeitschr. phys. Chemie 5*, 97–132.

_____. 1893. *Die Phasenregel und ihre Anwendungen.* Leipzig und Wien: Verl. Deuticke, 120 pp.

_____. 1895. Über reciproke Salzpaare. *Sitzungsber. Akad. Wiss: Wien* (1895): 840–55.

Naupert, A., & Wense, W. 1893. Über einige bemerkenswerthe Mineralvorkommnisse in den Salzlagern von Westeregeln. *Ber. Deutsch. Chem. Ges. 26*, 873–75.

Precht, H., & Wittgen, B. 1882. Löslichkeit von Salzgemischen der Salze der Alkalien und alkalischen Erden bei verschiedener Temperatur. *Ber. Deutsch. Chem. Ges. 15*, 1666–72.

Riecke, E. 1890. Beigräge zu der von Gibbs entworfenen Theorie der Zustandsänderungen eines aus einer Mehrzahl von Phasen bestehenden Systems. *Zeitschr. phys. Chemie. 6*, 268–80.

Rinne, F. 1911. Salzpetrographie und Metallographie im Dienste der Eruptivgesteinskunde. *Fortschr. Min. Krist. Petrogr. 1*, 181–220.

Roozeboom, H. W. Bakhuis. 1887*a*, Sur les différentes formes de l'équilibre chimique hétérogène. *Rec. Trav. Chim. Pays-Bas. 6*, 262–303.

_____. 1887*b*. Sur les points triples et multiples, envisagés comme points de transition. *Rec. Trav. Chim. Pays-Bas. 6*, 304–32.

_____. 1887*c*. Sur l'astrakanite et les sels doubles hydratés en général. *Rec. Trav. Chim. Pays-Bas. 6*, 333–55.

_____. 1888*a*. Studien über chemisches Gleichgewicht. *Zeitschr. phys. Chemie. 2*, 449–81.

_____. 1888*b*. Die Umwandlungstemperatur bei wasserhaltigen Doppelsalzen und ihre Löslichkeit. *Zeitschr. phys. Chemie 2*, 513–22.

_____. 1901. *Die heterogenen Gleichgewichte vom Standpunkt der Phasen-lehre*. Erstes Heft. Braunschweig: Vieweg und Sohn, 221 pp.

_____. 1904. *Die heterogenen Gleichgewichte vom Standpunkt der Phasen-lehre*. Zweites Heft. Braunschweig: Vieweg and Sohn, 457 pp.

Schreinemakers, F. A. H. 1891. Über das Gleichgewicht des Doppelsalzes von Jodblei und Jodkalium mit wässeriger Lösung. *Zeitschr. phys. Chemie 9*, 57–77.

_____. 1912. *Die heterogenen Gleichgewichte vom Standpunkt der Phasen-lehre*. Drittes Heft. Braunschweig: Vieweg and Sohn, 312 pp.

Sillén, L. G. 1959. The physical chemistry of sea water, *in*: Sears, Mary (ed.), *Oceanography*. *Amer. Assoc. Adv. Sci., Publ. 67*, 549–81.

Van der Heide, J. K. 1893. Die Doppelsalze von Kalium und Magnesium-sulfat: Schönit und Kaliumastrakanit. *Zeitschr. phys. Chemie 12*, 416–30.

Van Hise, C. R. 1903. Report on geophysics. *Carnegie Inst. Washington Year-book 2*, 173–84.

Van Rijn van Alkemade, A. C. 1893. Graphische Behandlung einiger thermo-dynamischen Probleme über Gleichgewichtszustände von Salzlösungen mit festen Phasen. *Zeitschr. phys. Chemie 11*, 289–327.

Van't Hoff, J. H. 1875. *La chimie dans l'éspace*. Rotterdam: Bazendijk, 44 pp.

_____. 1884. *Etudes de dynamique chimique*. Amsterdam: F. Muller, 215 pp.

_____. 1886*a*. Lois de l'équilibre chimique dans l'état dilué, gaseux ou dissous. *Kongl. Svenska Vet. Akad. Handl. 21*, 17–75.

_____. 1896*b*. *Studien zur chemischen Dynamik*. Leipzig: W. Englemann, 286 pp.

_____. 1897. *Vorlesungen über Bildung und Spaltung von Doppelsalzen*. Leipzig: W. Englemann, 95 pp.

_____. 1898. *Vorlesungen über theoretische und physikalische Chemie Erstes Heft: Die chemische Dynamik*. Braunschweig: Vieweg, 536 pp.

_____. 1902. Die Phasenlehre. *Ber. Deutsch. Chem. Ges. 35*, 4252–64.

_____. 1905. *Zur Bildung der ozeanischen Salzablagerungen*. 1. Heft. Braun-schweig: Vieweg and Sohn, 85 pp.

_____. 1909. *Zur Bildung der ozeanischen Salzlagerstätten*. 2. Heft Braun-schweig: Vieweg and Sohn, 90 pp.

_____. 1912. *Untersuchungen über die Bildungsverhältnisse der ozeanischen Salzablagerungen insbesondere des Stassfurter Salzlagers*. Precht and Cohen (eds.) Leipzig: Akad. Verl. Ges. 374 pp.

Van't Hoff, J. H., Armstrong, E. F., Hinrichsen, W., Weigert, F. & Just, G., 1903. Gips und Anhydrit. *Zeitschr. phys. Chemie 45*, 257–306.

Van't Hoff, J. H., & Van Deventer, C. M. 1887. Die Umwandlungstemperatur bei chemischer Zersetzung. *Zeitschr. phys. Chemie 1*, 165–82.

Walcott, C. D. 1901. *Annual Report of the Director of the U.S. Geol. Survey 22*, Washington, D.C., pp. 133–34.

_____. 1902. *Annual Report of the Director of the U.S. Geol. Survey 23*, Washington, D.C., p. 121.

_____. 1903. *Annual Report of the Director of the U.S. Geol. Survey 24*, Washington, D.C., p. 127.

_____. 1904. *Annual Report of the Director of the U.S. Geol. Survey 25*, Washington, D.C., pp. 102–3.

Chapter Three

The Odyssey of Geosyncline

KENNETH J. HSÜ

"I am doing the best I can at my age . . . "

G. B. Shaw, Preface to *Back to Methuselah*

INTRODUCTION

The concept of geosyncline was an American innovation by James Hall in 1859. The geosyncline was portrayed as a trough, concurrently filled with sediments. By the turn of the century this useful concept reached the fertile soil of European geology where it acquired new significance: A trough can be a geosyncline, especially if it is not filled up with sediments. Because there have been many different kinds of troughs, filled, or not filled, with different sequences of sediments, we have a kaleidoscope of geosynclines. Some are yours, some are mine; some are true, others not quite; most are gluttonous, a few are starved. Meanwhile, what constitutes a geosyncline has received no consensus. Several excellent summaries and historical reviews have been published during the last few decades (e.g., Knopf 1948; Glaessner & Teichert 1949; Aubouin 1965). So it might seem superfluous to return to this well-trodden debating ground. However, I have been prompted to choose this topic as my contribution to the Pettijohn volume by a seeming crisis in communication. We all know that the geosynclinal concept was conceived by a paleontologist studying ancient strata. The growth and evolution of this concept has come largely from interpretations by geologists as to the meaning of rocks. Aubouin (1965, p. 3) went so far as to state: "The concept of the geosyncline is a geological (i.e., a historical) concept which embraces not only the conditions of a given time but also their *prior* and *subsequent* record. For this reason, it is not possible to reject the concept, in favour of one founded on a knowledge of contemporary phenomena, gained either directly from oceanographic investigations or indirectly from geophysical data."

66

In taking his stand, Aubouin overlooked the fact that the strength of geology is derived from its uniformitarian basis, and that the geologic record is to be interpreted in terms of contemporary phenomena. Indeed, modern students demand actualistic models of sedimentation. It has been questioned if the geosynclinal concept, especially its special role in orogenesis, may not have outlived its usefulness. This sentiment was held by many during discussions at the Penrose Conference on New Global Tectonics, Monterey, 1969, and was summarized by Dickinson (1970, p. 22), the conferee: "By the close of conference, the trend of discussions had pointed clearly to difficulties with the traditional and valuable geosynclinal theory."

Should the geosynclinal concept be rejected? Has not the concept served a useful purpose? What are some of the pitfalls and traps of adhering to a mythical past? What have we learned and how? And where do we go from here? These are some of the questions that turned up when I was preparing this article. As an amateur historian, I would like to start this personal, perhaps not exactly impartial, account by following the threads that led us to where we are.

DOGMATIC SCHISM IN TWENTIETH CENTURY

The geosyncline, as implied epistomologically, means simply a large down-warped surface. To James Hall, this down-warping is related to the accumulation of an unusually thick sedimentary sequence, so that the line of maximum depression coincides with the line of maximum accumulation. Such a sequence in the Appalachian includes formations that contain abundant littoral faunas, which led Hall to conclude that geosynclinal sediments accumulated in shallow waters in environments similar to their coeval deposits on a tectonically stable shelf.

The European geologists, particularly the Austrian and French masters who studied the Alps before the turn of the century, took issue with Hall. Suess (1875, pp. 96–102) first advocated the idea that the geosynclinal sediments represent a pelagic facies. His arguments were, however, rather tenuous. He was impressed by the fact that geosynclinal sequences are not only thicker but also more complete and are devoid of unconformities, as compared to their shelf equivalents. He was wrong, however, to cite the Triassic of the eastern Alps, known to us as a tidal-flat complex (e.g., Fisher 1963), to prove his point that the alpine carbonates are largely pelagic. Equally erroneous is his interpretation that Mesozoic benthonic faunas of the Alps represented relic Paleozoic forms that had survived in oceanic deeps. Suess was evidently not a sedimentologist. In the same year, Neumayr (1875, p. 364, see also Neumayr & Suess 1920, pp. 369–71) was more successful with his actualistic approach to interpret the bathymetry of the alpine sediments: The Upper Jurassic and Lower Cretaceous radiolarian cherts of the Alps were compared to the Recent radiolarian oozes of the equatorial Pacific and Indian oceans. This interpretation has stood the test of time, and Neumayr's belief that the Tertiary radiolarian chert of the Barbados (Eocene Oceanic Formation) had been uplifted from a deep ocean bottom could be considered proven as a result of JOIDES drilling (Bader et al. 1970, p. 665).

The choice of the word "pelagic" by Suess permitted Walther in 1895 to attempt a conciliation of the American and European views on the bathymetry of geosynclinal sediments. In a paper on the biofacies of fossils, he popularized the terms "benthonic" and "planktonic," lately introduced by Hensen and Haeckel, and he emphasized that pelagic sediments containing planktonic radiolaria are not necessarily deposited in a deep-sea environment. In fact Walther (pp. 214, 237), referring to an expertise by John Murray of the *Challenger* Expedition, stated that the alpine radiolarites were not oceanic sediments.

Nevertheless, the Pandora's box of schism has been opened. A break-away from the Hall–Dana tradition in Europe was precipitated at the turn of the century with the publication of Emile Haug's masterpiece, *Les géosynclinaux et les Aires Continentales.*

Haug was born in Alsace in 1861. Like many others of his generation, he started his career as an ardent fossil collector in his youth. He studied at the provincial capital, Strassburg, and established himself as a young ammonite expert with his *summa cum laude* dissertation. This thesis was written in German, and he was required to write another in French when he became an exile in Paris, after some agitating activities as a radical student against the German annexation of Alsace. His childhood friend and countryman Kilian, who preceded him to the Sorbonne, was then working in the Maritime Alps, investigating mainly Late Jurassic and Cretaceous formations. Haug was prevailed upon to become a co-worker in the field, and applied his vast knowledge of ammonites to the difficult task of zoning the monotonous Jurassic black shales of the *dauphinois* Alps. In contrast to the shelf carbonate sequence in the Jura and Provence, the thick alpine or geosynclinal facies includes a series of continuously deposited sediments rich in ammonites, *Posidonomia*, and *Inoceramus*. However, it was not the difference in thickness, which so impressed Hall in New York, that caught Haug's attention. What was particularly noteworthy in the European example was the facies contrast which suggests two entirely different environments of deposition.

Haug, who introduced the terms "neritic" and "bathyal" in 1898, considered geosynclinal sediments the bathyal equivalent of neritic stable-shelf deposits, and he defined a bathyal water depth as ranging from 80 or 100 to 900 meters (Haug 1900, p. 620). His postulate was based upon: (1) the absence of unconformities within a geosyncylinal sequence; (2) the presence of pelagic fossils in geosynclinal sediments; and (3) the monotonous, shaly geosynclinal succession which suggests deposition at bathyal depth, in contrast to the shallow-water shelf-sediments which show great vertical and lateral facies variations in response to transgressions and regressions. Graptolitic shales, *Posidonomia* shale, *Aptychus* shale, and other formations containing pelagic fossils, have been cited by him as examples of typical geosynclinal deposits.

Haug's scheme has a charming simplicity and received an admirable boost from O. T. Jones (1938), who contrasted the *geosynclinal graptolite facies* with the stable-shelf *shelly facies* in the Paleozoic of Great Britain. Other British masters characterized geosynclinal sediments by their graywacke lithology (Tyrrell 1933), or by their typical sedimentary structure of graded bedding (Bailey 1936).

If nature had cooperated so that sediments *thicker* than their coeval sequences were always deposited in *deeper* waters, the new connotation by Haug would not lead to any difficulty. Unfortunately, this is not the case. Take the Middle and Upper Devonian of the eastern United States, for example; the thin Chattanooga shale of the Interior Platform was deposited in much deeper waters than its coeval deltaic Catskill and pro-deltaic Chemung formations of the Appalachian geosyncline.

Haug's scheme led to a crisis: one had to adhere either to Hall's original definition or to select bathymetry as a new criterion. The Americans stood by Hall, the Europeans followed Haug, resulting in the so-called "American" and "European" concepts of geosynclines (see Trümpy 1960, p. 865; Aubouin 1965, p. 17). To conform with the European concept a "lepto-geosyncline" had to be fabricated to designate deep-sea deposits which are thinner than their shelf equivalents, such as the Upper Jurassic and Lower Cretaceous pelagic deposits of the Austroalpine and Pennine Alps. Recognition of the leptogeosynclinal nature of a formation depends on the evidence indicating deposition at bathyal or greater depth (Trümpy 1960, p. 865). If we take this statement at its face value, we would come to the paradoxical conclusion that the Devonian geosyncline in eastern North America was not located in the Appalachian Mountains, but in the Interior Platform, on which the deep-water Chattanooga shale was deposited. This interpretation is certainly not what James Hall had intended when he first proposed the geosynclinal concept, nor would Haug or Trümpy have come to such an extreme view.

MOBILITY AND GEOSYNCLINAL CYCLES

We have seen the lack of consensus on the definition of geosyncline. Indeed, does the word have any meaning, and why do we choose to use such a word?

There is one common denominator underlying all the different schools of thought: Geosyncline implies mobility and this mobility is represented by an unusually high subsidence rate, giving rise to either unusually thick sequences or unusually deep water deposits. The original concept proposed by Hall and developed by Dana was designed to emphasize the existence of such mobile zones within the earth's crust. Haug (1900), Schuchert (1923), and Bucher

(1933) all portrayed geosynclines as long linear belts, mediterranean, or marginal to continents. Under different geographical settings, this mobility has found manifold expressions, resulting in the wide varieties of sedimentary prisms. The courageous, though not necessarily gratifying, attempts by Stille (1941) and Kay (1951) to characterize those various sequences led eventually to the edifice of a complicated and controversial nomenclature. So we have orthogeosynclines and parageosynclines; eugeosynclines and miogeosynclines; exogeosynclines, autogeosynclines, and zeugogeosynclines; taphrogeosynclines and epieugeosynclines; deltageosynclines and paraliageosynclines, etc., etc.

The Stille–Kay nomenclature emphasized the spatial variation of the geosynclinal mobility. The next question refers to its temporal development. Have there been any variations in the tempo and mood of this mobility at any given belt? Could we detect any pattern or rhythm in the sedimentary sequence of a geosyncline? Marcel Bertrand was one of the first to find such a rhythm, through his recognition of the flysch-molasse succession.

Marcel Bertrand could be considered one of the rare geniuses in geology. As a theoretician, he correctly interpreted the nappe-structure of the Glarus Alps, while reading Suess's *Die Entstehung der Alpen* on a train. Yet his logical mind, which seemed to have always managed to sort out order from apparent chaos, may well have been the culprit that drove him to apathy in life after the irrational loss of his only child. He delivered his swan song at the Zurich International Geological Congress in 1894; his messages were short and clear: he innovated the concept of recurrent facies, established the principle of geosynclinal cycles, and presented evidence for the ensialic origin of geosynclines.

Bertrand (p. 170) recognized four "facies" or "*formations de montagne*" which constitute a geosynclinal cycle (see Figure 1):

A. *Gneiss*, which constitutes the basement of a primary geosyncline.

B. *Flysch schisteux* (= *schistes lustres*), which filled up the central zone of the primary geosyncline.

C. *Flysch grossier* (= *flysch*), which filled up the border of the geosyncline after the elevation of the central axis.

D. *Grès rouges* (= *molasse*), which was developed at the foot of the newly elevated mountain chain.

The subsequent development on this theme mirrored the trend of the sedimentological research in the twentieth century. In the first two decades, when the embryontectonics (Argand 1916) was new and fashionable, the tendency was to relate sedimentation to geosynclinal growth. Arbenz (1919) recognized an "epeirogenic facies" of cyclothemic deposits in the Helvetic Alps, followed by an "orogenic facies" of flysch and molasse. Those facies are not defined on the basis of their sedimentary environments, but are manifestations of the tempo and mood of tectonic mobility.

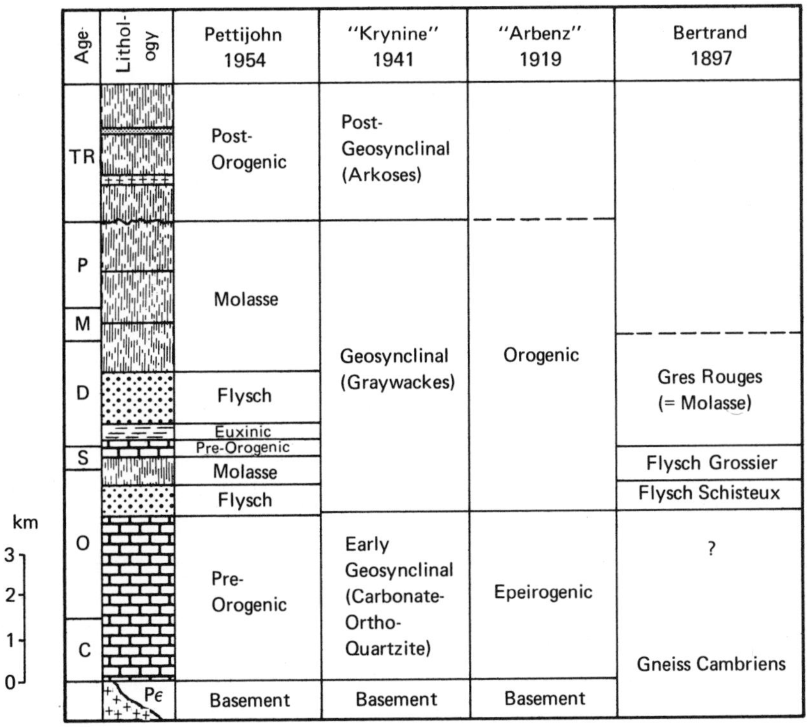

Age	Lithol-ogy		Pettijohn 1954	"Krynine" 1941	"Arbenz" 1919	Bertrand 1897
TR			Post-Orogenic	Post-Geosynclinal (Arkoses)		
P			Molasse			
M				Geosynclinal (Graywackes)	Orogenic	
D			Flysch			Gres Rouges (= Molasse)
S			Euxinic Pre-Orogenic Molasse			Flysch Grossier
			Flysch			Flysch Schisteux
O			Pre-Orogenic	Early Geosynclinal (Carbonate-Ortho-Quartzite)	Epeirogenic	?
C						Gneiss Cambriens
		Pє	Basement	Basement	Basement	

km
3
2
1
0

Fig. 1. Evolution of the concept of geosynclinal cycles.

The terms "flysch" and "molasse" and the concept "orogenic facies" were brought across the Atlantic and rooted in North American literature by Waterschoot van der Gracht in 1931. He described flysch "as sediments deposited previous to the major paroxysm" and molasse as those deposited "during or immediately after the major diatrophism" (p. 998). The preoccupation of geologists with orogenic chronology in the following two decades (e.g., Stille 1924, 1936; Bucher 1933; Gilluly 1949) may have resulted in the overemphasis of the temporal relationship between flysch, molasse, and orogeny. Adding pre-, syn-, and post- to the term "orogenic facies" was predestined to cause confusion as soon as the "one big-bang" concept of orogeny was proven unfounded.

We saw the growth of sedimentary petrography during the twenties, helped in part by the industry's effort to use mineral-zonation for correlation. Eventually the chronostratigraphical values of minerals proved limited, yet the tectonic connotations of sediment composition did not escape attention. In his presidential address to the Geological Society of London, O. T. Jones (1938) attempted a petrographical characterization of geosyncline, and graywacke was singled out as its essential deposit. This approach was much appreciated on the other side of the Atlantic, where Krynine (1941), Pettijohn (1943), Krumbein, Sloss, and Dapples (1949), and others have all joined in to develop the theme on the significance of sedimentary petrology.

We still remember Paul Krynine as a colorful personality. Unfortunately, he left behind relatively few publications for posterity. Nevertheless, apparently in a blaze of creative frenzy, Krynine presented his doctrines in eight abstracts for the Boston meeting of the Geological Society of America in 1940. The first of the eight, "Differentiation of sediments during the life history of a landmass," contained the essence of his scheme (1941, p. 1915). He divided geosynclinal cycles into three stages (see Figure 1): "(1) Peneplanation (or early geosynclinal) stage: cyclic deposition on fluctuation flat surface after much weathering, characterized by *first-cycle quartites* . . . (2) Geosynclinal stage: basinal deposition interrupted by local vertical buckling. . . . Typical sediment: *graywackes* . . . (3) Post-geosynclinal stage: uplift . . . after folding and magmatic intrusion of geosyncline. Typical sediments: *arkoses.*" Krynine found his model in the Appalachians. Yet the situation in the Alps was quite different: there the "geosynclinal" flysch sediments are arkosic, having been derived from an old granite landmass, whereas the "post-orogenic" molasse detritus are largely rock fragments that came from the cover-nappes topping the nearby newly elevated mountains.

The mineralogical composition is certainly an important attribute of a rock; "the presence or absence of a given mineral may be an important clue to the history of the rock" (Pettijohn 1957, p. 107). On the other hand the mineralogy depends not only upon tectonic mobility but also upon too many other complex factors to be an infallible indicator of geosynclinal development. This fact had been generally recognized by 1960, so I was whipping a dead horse when I published my "new" discovery on the mineralogical significance of Recent sands that year.

In 1930 Sir E. B. Bailey made an observation that there are two principal types of internal structure in sandstones, namely, cross-bedded and graded. He noted that these structures are more or less mutually exclusive and may characterize two contrasting sedimentary facies. The significance of sedimentary structures has not been overlooked by other Europeans. Brinkman (1933) and Cloos (1938) in Germany, and Migliorini (1943) in Italy all published on the paleogeographical implications of those structures. In the United States Pettijohn was among the first to discern the importance of the work by Jones and Bailey. He (1943) related Bailey's "graded facies," and Jones's "graywacke suite" to the alpine "flysch-type" sediments. Through his investigations of the Archaean rocks of the Lake Superior region, he further noted a kinship between flysch and eugeosynclinal sedimentation.

A new era of sedimentological research dawned after World War II. Knight's (1929) and Reiche's (1938), pioneer works remained largely unnoticed, until the search for uranium in the Colorado Plateau country led to a blossoming of studies of cross-bedded rocks (e.g., Stokes 1947; McKee & Weir 1953). At about the same time, Migliorini's cries in the wilderness received experimental support from Kuenen and the publication of their joint paper in 1950, as foreseen by Pettijohn (1950), opened the floodgate of flysch research. Meanwhile, Pettijohn was playing a most important role in this sedimentological break-through. His work and that of his students of The Johns Hopkins University in the Appalachians has spearheaded an entirely new approach to basin analysis. Furthermore, their carefully compiled data permitted a new appraisal of the concept of the geosynclinal-cycle.

Pettijohn (1957, p. 639) emphasized that the main influence of tectonism "is the effect on the total sediment supply and the rate of subsidence." Following Krynine, he recognized a pregeosynclinal (or early geosynclinal) stage of orthoquartzite-carbonate sedimentation. Euxinic facies or flysch facies followed, as tectonic mobility accelerated. The sedimentary record is then "essentially a record of basin filling" (p. 640), culminating in molasse deposition. This analytical approach enabled him to perceive a polycyclic geosynclinal development in the Appalachians, which could be related to the multiple orogenic events there.

Figure 1 is a graphical summary of the development of our thinking on the geosynclinal cycle. The Appalachian sedimentary sequence of the Ridge and Valley province serves as an illustration. Bertrand referred to this section in his 1897 paper; his facies interpretation, based upon reading of the literature, necessarily contained some grave errors, but he did grasp the essence of geosynclinal development. Arbenz did not actually discuss the Appalachians in his 1919 article, but he probably would have come to the conclusion that I have presented here in his name had he visited North America. The column styled after Krynine, I believe, is an accurate portrayal of his scheme. The last column is taken from the book by Pettijohn (1957, p. 641). Some may take issue with him on the names of those consanguineous associations, but I saw no signs during the last decade that his interpretation could be altered. Here is one of the many lasting contributions made by the man we honor by this volume.

ISOSTASY AND GEOSYNCLINAL SUBSIDENCE

Hall (1859) thought geosynclinal subsidence was induced by sedimentary load. Dana (1873) thought the trough was compressional in origin. After the concept of isostasy was introduced by Dutton (1889) the question of whether or not the sinking is isostatic became a focus for speculation during the next seventy years.

Assuming that a sedimentary column cannot be built above sea leavel, Jeffreys (1924, p. 143) made a simple calculation that "the maximum depth of the sediments is 2.2 times the original depth of the water," if isostatic equilibrium is *steadily* maintained. Bucher (1933) made a similar estimate on the basis of somewhat different assumptions of crust and mantle densities. Nevertheless, geomorphic evidence in the Mississippi delta region led Russell (1940) and Fisk (1952) to revive the idea of load-subsidence. So Lawson (1942) carried out an arithmetical manipulation again; he showed that the primeval trough had to be 16,000 feet deep to account for the accumulation of deltaic sediments there. Since few Americans at that time believed in the existence of deep-water sediments in geosynclines, Lawson's computation could be regarded as a conclusive argument to rule out isostatic subsidence.

One of the most conclusive pieces of evidence on the deep-water origin of those sands which are now called turbidites was presented by Manly Natland in 1933. Years later, when I was working in Ventura, I heard stories of Natland's excitement and joy when he found practically identical foraminiferal assemblages in the Pliocene sediments of Ventura Avenue Field as those living today offshore in deep-water basins; and of his disappointment and frustration when neither his close associates nor the profession as a whole were ready for his outrageous hypothesis that the sands and conglomerates of the Ventura Basin were deposited in waters thousands of feet deep. After the war, E. L. Winterer started field work in Ventura and was among the first to find physical evidence to support Natland's faunal interpretation, but his monograph (with Durham 1962) was not published until more than ten years later because of governmental bureaucracy. Meanwhile, in 1950, Natland (with Kuenen) was able to find at an SEPM symposium the receptive audience to his heretofore neglected idea of deep-water sedimentation.

That a geosynclinal prism includes deep-water sediments and the recognition of flysch-molasse succession gave a new lease on life to the isostatic hypothesis. Furthermore, it occurred to me that all previous computations on the depth of the "primeval trough" forgot the role of sediment-supply, nor did they take compactional subsidence into consideration. My 1958 paper was an attempt to provide a theoretical basis for the geosynclinal cycle recognized by Pettijohn; the theory demands that the initial stage be the starved euxinic facies, followed by filling flysch and molasse, and should end in carbonate and orthoquartzite, when the limit of isostatic subsidence is reached.

In 1955, when the *Crust of the Earth Symposium* was published, Airy's model of isostasy reached its zenith of popularity: The water-depth of earth's depressions was related directly to the thickness of underlying crust (Heiskanen & Vening Meinesz 1958). It seemed clear that the crust underlying a geosyncline must have been unusually thin during the stage of flysch sedimentation (Hsü 1958). Because geosynclines such as the Appalachian or the alpine have undergone stages of transition from carbonate-orthoquartzite to euxinic or flysch sedimentation (Pettijohn 1957; Trümpy 1970), the more relevant question was now not subsidence under load, but the cause of crustal thinning.

Adhering to Airy's model of isostasy, there were only two possibilities: either the crust is thinned from below, thereby inducing isostatic subsidence; or a thicker crust has been rifted apart, causing the median valley to be floored by a thin basaltic crust.

There are several variations on the subcrustal thinning theme. Gilluly (1955) introduced rather casually the concept of "subcrustal erosion." Kennedy (1959) and others favored the idea of phase change from a basalt crust to an eclogite mantle. Van Bemmelen (1958) and Beloussov (1962) speculated on chemical basification of the crust.

The rifting hypothesis is a corollary of continental drift. Carey (1958) introduced sphenochasm and rhombochasm. With the gradual acceptance of sea-floor spreading and continental-drift theories by the geological community, it became almost self-evident that the subsidence of a stable continental margin can be related to tensional faulting in the early stage of rifting (e.g., Heezen 1969).

A third alternative of supracrustal thinning seemed impossible as long as the Airy's isostatic model remained supreme. However, even in 1955 all the pieces did not fit into the model (Tatel & Tuve 1955). By the early sixties it became increasingly apparent that lateral variation of mantle density exists (Pakiser 1963). Two long forgotten terms "lithosphere" and "asthenosphere," first introduced by Barrell in 1915, were exhumed to replace crust and mantle as designations for the outer shell of strength and the underlying zone of weakness.

Since lithosphere includes both crust and mantle, the surface elevation of an isostatically adjusted column depends not only on crustal thickness but also on mantle density. This formed the basis of my postulate of supracrustal removal as the cause of crustal thinning (Hsü 1965). This idea that erosion and/or tectonical denudation are the main causes of the disappearance of former continental areas has been further developed by Schuiling (1969) to explain the foundering of such complex basins as the western Mediterranean or the Black Sea.

Whereas crustal thinning and mantle-density changes had to be invoked to explain the origin of ensialic geosynclines, we all recognized that sedimentary load can still be the primary cause of subsidence, if the floor of the geosynclinal is ensimatic (Hsü 1958; Dietz 1963). This leads us to the next question: Are there any ensimatic geosynclines?

ENSIMATIC AND ENSIALIC GEOSYNCLINES

Having grown up in a Mediterranean land, Haug (1900) envisioned geosynclines as long, narrow, intercontinental troughs. He soon ran into difficulties when he looked toward the Far East. To account for the circum-Pacific mountains, he had to assume a lost Pacific continent. It is interesting to note that the idea of sunken Pacific was taken seriously even in the sixties by as great an authority as Beloussov (with Rudich 1961). Others are less gullible.

When I read again the first continental-drift symposium volume (van der Gracht 1928), it occurred to me that the fierce opposition Wegener (1912, 1922) met in the New World in contrast to the support he received in Europe might reflect the different schools of thinking on the nature of geosynclines. The descendants of Hall and Dana had no place for oceanic sediments in geosynclines. Meanwhile, on the other side of the Atlantic, Kossmat (1921), Argand (1922), and Staub (1928) found a ready explanation for the bathyal facies they saw in the Alps by rifting continents and creating intercontinental oceanic basins.

As a beginning graduate I studied for my language examination by reading Staub's *Der Bewegungsmechanismus der Erde*. Staub's work has fallen into oblivion probably because of his unwise appeal to the *polflucht* ("flight from the pole") as the motivating force. This urge might indeed induce the mass migration of Swedes and Germans to the Mediterranean beaches, but it is certainly not strong enough to account for the wandering of continents (Griggs 1939). Yet Staub's book, if redecorated with modern jargons, would serve as an excellent text for students of new global tectonics. He had enough foresight to see that the alpine rocks belong to two fundamental types: those deposited on subsided continental margins, and those in an oceanic geosyncline (Staub 1928, p. 9). Some twenty years later they would be christened ensialic and ensimatic geosynclines by Wells (1949).

In a concise but thought-provoking essay prepared by Jim Gilluly for the *Crust of the Earth Symposium* the question of ensimatic geosyncline was explored. Gilluly (1955) came to the somewhat startling conclusion that all major geosynclines are ensialic, except perhaps the Franciscan of the Pacific coast ranges. Geological evidence seemed overwhelming in support of this view, even the ophiolite of the Alps appeared to have been emplaced on a sialic basement (Trümpy 1970).

I myself had an opportunity to examine some of the evidence. For example, blocks of granite and Triassic shelf carbonates were found in the ophiolite zone of Oberhalbstein in Grisons, Switzerland, in association with Jurassic and younger blocks detached from the Steinmann Trinity; all are embedded in a pervasively sheared matrix. If all those constituted a normal stratigraphical succession, we must conclude that the ophiolite had an ensialic basement. However, the alpine ophiolite zones consist largely of tectonic mélanges. Tectonically mixed blocks may have been derived from widely separated paleogeographical realms (Hsü 1968 1971); thus the presence of granite in an ophiolite mélange does not yield irrefutable evidence of an ensialic geosyncline.

Do we have evidence that eugeosynclines are ensimatic? Before this question is answered, we should examine the definition: The term "eugeosyncline," proposed by Stille (1941), has been defined as a "surface that has been subsided deeply in a belt having active volcanism" (Kay 1951, p. 4). Do such surfaces exist, and have they ever existed? And what is the nature of the concurrent volcanism? These questions bring us to the next chapter, which reviews the ideas relating magmatism to geosynclinal development.

MAGMATISM AND EUGEOSYNCLINE

The term "ophiolite" was used in the last century by Brongniart as a Greek-derived synonym for "serpentinite" (Stille 1941, p. 14). Steinmann (1905, 1926) broadened the meaning of the term to include not only this ultramafic rock but also associated ultramafic and mafic rocks, such as peridotite, pyroxenites, gabbro, diabase, spilite, etc. They constitute the ophiolite suite.

In 1905 Steinmann recognized the association of ophiolites with radiolarian cherts and other deep-sea sediments. The association (radiolarite-greenstone-serpentinite) was eventually given the imposing title of the "Steinmann Trinity" by Sir E. B. Bailey (Bailey & McCallien 1960). Suess (1909, p. 645) reported the occurrence of ophiolites in the Appalachian Piedmont and in other mountain belts and remarked that they never occur in foreland regions. He also recorded the finding of submarine basalt from the Atlantic, of gabbro in Iceland, and of peridotite on St. Paul's Rock. Suess agreed fully with Steinmann that the ophiolites were associated with deep-sea deposits; he noted, however, that they might acquire strange bedfellows through tectonic mixing (p. 646).

Both Steinmann and Suess recognized that alpine ophiolites represent raised ocean floors. Kossmat was similarly inclined when he first evolved the idea of igneous cycles in 1921. Other German masters, particularly Scheumann (1932), von Bubnoff (1937), and Stille (1941), deviated from this view when they elaborated on the scheme of the pre-orogenic, synorogenic, and post-orogenic stages of magmatism. With the creeping "fixistic" doctrine of the forties, ophiolites were no longer referred to as evidence of a simatic crust. Instead, they were considered the earliest phase of magmatism of an ensialic geosyncline, when the sialic crust was not yet buried deep enough to yield magmas. Only later, during synorogenic compression, the temperature at the base of the subsided and thickened crust was raised high enough to produce granite magmas (Stille 1941; Knopf 1948).

The transplanting of the eugeosyncline concept on American soil was a source of further confusion: all European geologists emphasized the importance of ophiolites in "pre-orogenic magmatism", but the American importers did not make such a fine distinction on the nature of volcanics. Knopf (1948) cited thick rhyolite sequence as evidence for "the role of volcanism . . . during the growth of the preparatory geosyncline." Kay (1951) rendered the "Magog Belt" eugeosynclinal even though the diversified volcanic rocks there include rhyolite, dacite, andesite, pyroclastics, and other non-ophiolitic rocks. The inclusion of andesitic flows as a major component of an eugeosynclinal package led naturally to the image that modern trenches and island arcs represent the actualistic setting for ancient eugeosynclines (Kay 1951).

A still more confusing development is the controversy concerning the significance of spilite. Near the turn of the century Harker (1896, 1909) and Becke (1903) popularized the terms "Atlantic suite" and "Pacific suite" for saturated and undersaturated (with respect to silica) basalts. In 1911 Dewey and Flett discovered the high soda content of some pillow lavas associated with geosynclinal sediments; they proposed a "spilitic suite" as the third major type of basalt, typically geosynclinal. Thus a controversy was born: Is spilite primary, or does it simply represent albitized basalt? Gilluly's (1935) work on Oregon spilites has practically settled the question in favor of the latter view. However, a second problem remained unresolved: Is this alteration a result of regional green-schist metamorphism, or is it associated with magmatic or hydrothermal activities during emplacement?

W. Q. Kennedy reviewed the basalt problem in 1938. By then the geographical designations "Atlantic" and "Pacific" had proven inadequate and Kennedy proposed substitute terms characterized by their chemistry, namely, "alkaline" and "calc-alkaline." The latter was also known by the now popular name of "tholeiitic magma-type." Kennedy's article established the petrological essence of these two basalt suites; spilite found no place in this scheme. However, he erred when he wrote that "the tholeiitic magma is consistently absent from (ocean basins)" and that tholeiite "appears to be connected with the granitic crust" (Kennedy & Anderson 1936, p. 38). This mistake was explicitly corrected by Tilley in 1950 (p. 40), when he delivered his presidential address, "Some aspects of magmatic evolution"; we now know that ocean basalts are largely tholeiitic (Engel & Engel 1964).

The ophiolite question and the spilite puzzle were finally resolved during the last decade. On land, detailed petrological and geochemical analyses of eugeosynclinal rocks indicated that pillow basalts are not all spilitic and that spilites could be considered altered tholeiites (e.g., Bailey et al. 1955; Moores 1969). Under the sea, pillow lavas consisting of normal tholeiites have been hauled up repeatedly (e.g., Moore 1965). Altered tholeiites that can be called spilite, as well as plutonic members of the ophiolite suite, have also been dredged from oceanic ridges (e.g., Melson & van Andel 1966; Cann 1967; Quon & Ehlers 1963).

Thus ·it was not surprising that we are now returning to the idea expressed at the beginning of this century by Steinmann and Suess: a consensus was voiced during the discussions at the Penrose Conference, Monterey, 1969, that alpine ophiolites are raised ocean floors (Dickinson 1970). This interpretation received confirming support when deep-sea drilling southwest of Lisbon penetrated a typically alpine Mesozoic sequence before coring its ophiolite basement, which consists of spilite, gabbro, and serpentinite (Ryan et al. 1970). There we found a relic of the alpine ophiolite-geosyncline still existent as a part of an ocean basin; its ensimatic origin can hardly be questioned.

During the last decade it also became increasingly apparent that an indiscriminate reference to all volcanics as a part of eugeosynclinal sequence has led to confusion. The origin of andesitic volcanics is very different from that of the ophiolites. Andesite occurs in abundance in circum-Pacific island arcs and coastal mountains, but rarely within the Pacific Basin inside the andesite line, where basalts are found. Lately andesitic volcanism has been considered the surficial equivalent of batholithic activities (e.g., Hamilton 1969), and the geneses of both have been related to the partial melting of a lithospheric plate that descended behind an island arc or a trench margin (e.g., Dickinson & Hatherton 1967; Hamilton 1969). A eugeosyncline could indeed by ensialic if the presence of andesitic volcanics is used as a criterion to distinguish it from a miogeosyncline. But then the term "eugeosyncline" would be meaningless. The ophiolite magmatism that takes place under a mid-ocean ridge should not be placed in the same category as the granite- and andesite-magmatism behind a trench margin.

GEOSYNCLINAL THEORY OF OROGENY

When Hall innovated his geosynclinal concept in 1859, he presented a theory of orogeny: Sedimentary load led to subsidence, which in turn led to the folding of geosynclinal sediments and orogeny. Dana (1873) placed a different emphasis. The first cause was compression, which formed troughs as sites of sedimentary accumulation and led ultimately to the formation of mountains. Yet both of those American masters assumed geosynclines to be precursors of orogeny. Again Suess (1875) elected to differ. He found that sedimentary sequences in a mountain chain are not necessarily "geosynclinal" (p. 120) and that pelagic sediments, which were considered geosynclinal by him, are not uncommon on stable platforms.

Eduard Suess was the author of *Das Antlitz der Erde*, which remains a milestone as yet unequaled. Many of our fundamental concepts in geology can be traced back to this masterpiece: Sial and sima; Atlantic and Pacific types of continental margin; Tethys and Gondwanaland. Suess was the professor of geology at the University of Vienna during the second half of the nineteenth century, when Vienna was a foremost center of European culture.

His students were to occupy chairs in many universities, and through them he left a strong influence on the development of geology in Europe.

I combed through the thick autobiography by Suess to look for clues that might betray his scientific temperament. Son of a German protestant minister and married to a Jewish banking family of the Austro-Hungarian empire, Suess was eighteen when the Revolution of 1848 erupted. Contrary to our present turmoil, when radicals excell in oratorical brilliance, the generation of 1848 were activists and those in Vienna carried out their reform with deeds. As a cadet of the "Student Corp," Suess was given a rifle to guard the basement of a bank against vandals. Throughout his life Suess was a democrat in an aristocratic society. He was active in community affairs and later in national politics, always seeking to initiate programs to benefit his countrymen in particular and mankind at large. Yet he angrily resigned from the city legislature when it sought to raise funds for one of his own pet projects through a somewhat dubious means (lottery). Perhaps that was the clue I sought: Suess was an uncompromising man of principle who would not overlook any factual details that might blemish a beautiful theory. This master of old global tectonics was never seduced by a great inductive generalization, which, as Adolf Knopf (1948, p. 649) put it, "was based on a single sample, the Appalachians."

While Suess's iconclastic pronouncement against Hall on the depositional environments of geosynclinal sediments was to form the foundation of a European school, his skepticism of the geosynclinal theory of orogeny found few adherents in Europe. Ironically, on this issue some American geologists were to take Suess's position, like Harry Hess (1951, p. 529), who argued that "there is no good reason to suppose that a geosyncline will locate future mountain building," or Marshall Kay (1951) who spoke of post-orogenic taphrogeosynclines or of cratonic autogeosynclines. Nevertheless, barring such exceptions, the main stream of geological thinking on both sides of the Atlantic carried on in the Hall–Dana tradition. The conviction was expressed by Walter Bucher (1933, p. 126) as one of the geological "laws": "Law 20: The typical orogenic cycle begins with geosynclinal depression and ends with a major uplift. The interval between these limiting events comprises two phases. The first phase is one of quiet sinking, only occasionally interrupted by uplifts; the second phase consists of crustal foldings separated by diminishing epochs of geosynclinal sinking."

The concept was so deeply ingrained that when Dave Griggs (1939, p. 314) looked for some generalized patterns of mountain-building he did not hesitate to express the "consensus of geological opinion that the history of all the major mountain systems has followed this broad outline of events."

However, the geosynclinal theory of orogeny was more a creed of the European and eastern establishments, the dogma never found ready acceptance among the rugged individualists of the West, many of whom had to deal with the reality of applying geology to oil-finding. The situation was particularly unorthodox in southern California: Where was the preparatory geosyncline? When did the orogeny reach its paroxysm?

One of the very tough questions in the qualifying examination prior to my admission to Ph.D. candidacy was to write an essay on whether or not the Ventura Basin had gone through a geosynclinal stage. I could have resorted to a semantic trick and avoided the issue through a positive reply by naming the thick and nonvolcanic sequence there epi-eugeosyncline. But then I would be ignoring the issue that such a "geosyncline" had not behaved as the theory predicts: Where was the pre-orogenic volcanism? Where was the syn-orogenic batholith?

Years later I visited the Ventura Basin with a venerated Swiss geologist trained in the best alpine tradition, we argued over the question of whether the graded beds of the Ventura Basin were flysch or molasse. Those sediments are flysch sedimentologically. However, my friend emphasized, not without some justification, that they were molasse deposited in "post-orogenic" basins. To him, flysch recorded sedimentation in the foredeep of an embryonically folded geosyncline. Since the Ventura sediments were not geosynclinal, they were not flysch, no matter what their sedimentological character indicated.

Later a local expert on California geology, A. O. Woodford (1960), came to a similar conclusion, that the geosynclinal stage and the major orogeny took place in southern California during the Mesozoic; the Cenozoic sediments there constitute merely the "superjacent series." Even I myself was inclined to this view, when I speculated on the relation between the Franciscan and the younger rocks (Hsü 1967). So those thick sediments in local basins should not be considered geosynclinal. Yet have they not been orogenically deformed?

I remember vividly my argument with my Swiss friend at Ventura, as we stood in front of a tightly folded outcrop of the Ventura anticline. He insisted that the Ventura sediments are post-orogenic, even when I pointed out to him the severe deformation to which those turbidites have been subjected. Nonetheless, he maintained what we saw represented no orogenic deformation because *there were no nappes.* To him all the folding and thrusting in southern California are merely insignificant local events which could not be construed as evidence of orogeny!

Not all of us can subscribe to such an extreme view. To us who lived in the Wild West, alpine deformation is only one of the several expressions of orogeny. After all, the term was originally coined by the pioneer G. K. Gilbert (1899) to designate the faulting along the Wasatch Front. Adhering to this view, movements along large faults, such as the San Andreas, involving hundreds of miles of lateral slip and creating mountains thousands of feet high are certainly orogenic. Yet one finds no preparatory geosynclines along such faults.

In fact, not all alpinists are equally enchanted with the idea that geosynclines are precursors of nappe deformation. Gansser (1965) remarked that the folding of the mightiest range in the world, the Himalayas, was not preceded by a preparatory geosynclinal stage. In this pronouncement Gansser was probably motivated by the observation that the bulk of the Himalaya sediments belong largely to the "foreland" facies, but not to the "geosynclinal" facies as defined by Haug. With the popularization of the sea-floor spreading and plate-tectonics, the geosynclinal theory began to seem superfluous. Even an alpine authority was moved to state that "the classical concept of 'geosyncline' is incompatible with the abundant recent information" (Laubscher 1969). It became increasingly apparent that orogeny affects mostly the sediments of plate-margins in sinks of global lithospheric movements. In fact such plate-margins need not be covered by a thick sedimentary column to localize orogenic deformation, as deep-sea drilling on Gorringe Bank in the eastern Atlantic indicated. There the ocean floor of the African plate, veneered with thin sediments, was overthrust onto the European plate (Hsü & Ryan 1971). Furthermore, thick sedimentary sequences accumulating on a continental margin within a lithospheric plate, such as those on the Atlantic margin of North America, may reach geosynclinal proportions but remain undeformed. These observations fully justified Suess's skepticism about the causal relation between geosyncline and orogeny.

GEOSYNCLINE AND PLATE TECTONICS

Before concluding my essay I would like to pose the questions again: What is a geosyncline? Do geosynclines exist?

I tend to think that geosynclines as they are described in some classical writings can be semantic traps. Take the case of the eastern continental margin of North America as an example. Figure 2 shows the crustal section as determined by geophysical surveys and by drilling. Where is our geosyncline?

Using the thickness as a criterion, one would not hesitate to state that the thick Bahama–Florida carbonates constitute a typically miogeosynclinal sequence. Yet is the surface underlying this sequence synclinal?

Using the sedimentary facies as a criterion, one would incline to view the thin sequence above the basalt east of the Bahama Platform as eugeosynclinal. But where is the "surface that has subsided deeply in a belt having active volcanism"? The ocean floor east of the platform has remained largely stationary since Jurassic and was not a site having active volcanism. The belt having active volcanism is the Mid-Atlantic Ridge, thousands of miles east of the Bahama Basin.

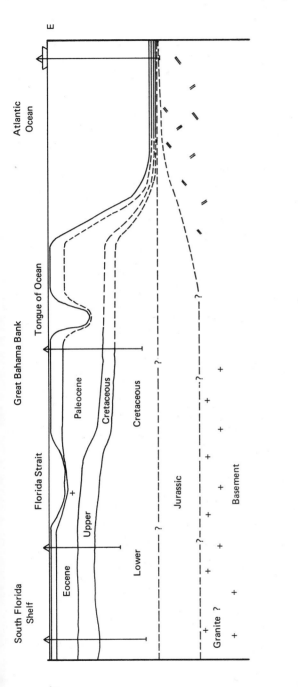

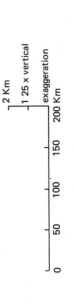

Fig. 2. Crustal section from the Florida–Bahama area, based on Sheridan et al. (1966), and Ewing et al. (1969).

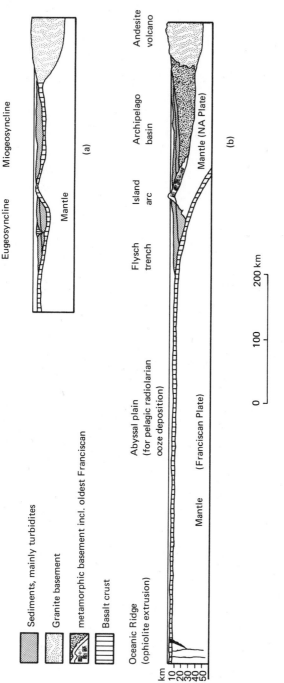

Fig. 3. Reconstruction of the Franciscan eugeosyncline: (a) as the outer member of a eu-miogeosynclinal couple, traditional model (Bailey & Blake 1969), (b) as a segment of the Pacific Ocean, plate-tectonic model (Hsü 1971).

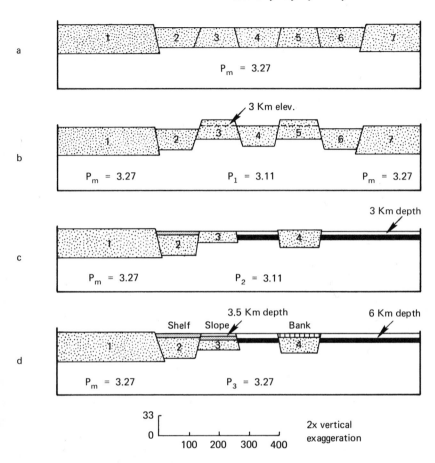

Fig. 4. Supracrustal thinning and mantle density as the primary cause of the subsidence of continental margins. Note the block no. 3 which was uplifted during the initial stage of rifting to provide clastic sediments for rift valleys. This block of thinned crust would subside eventually down to 3 to 5 km depth to form the outer continental margin of block no. 1 (N. America). Block no. 4 forms shallow marine banks, such as the Bahamas. Block no. 7 drifted away as another continent (Europe). For a detailed quantitative analysis of supracrustal thinning, mantle-density changes, and isostatic subsidence, see Hsü (1965).

In a recent paper on "Franciscan mélange as a model for eugeosynclinal sedimentation and underthrusting tectonics" (Hsü 1971), I challenged the traditional view which represented eugeosynclines as elongated troughs. The so-called eugeosynclinal rocks of the California coast ranges represent tectonic mixtures of rocks derived from three realms of deposition. The Franciscan mélange includes ophiolites emplaced at a mid-ocean ridge, radiolarites deposited on an abyssal plain, and flysch turbidites poured into a trench on the continental margin. Thus the Franciscan eugeosynclinal realm included much of the northwest Pacific, thousands of kilometers wide, as represented by Figure 3*b*. The idea of parallel troughs, constituting mio- and eu-geosynclinal couplet, visualized by Aubouin and adopted by Bailey and Blake (1969) for the Franciscan (Figure 3*a*) may constitute a grossly misleading representation.

Oceanographical explorations during the last decades have also helped dispell the myth about miogeosynclines. Those are apparently sediments deposited on stable continental margins. Combining the rifting hypothesis with supracrustal erosion, the crustal thinning along continental margins seems to represent a consequence of continental drift as indicated by Figure 4. The cause of geosynclinal subsidence would then be readily explained in terms of mantle-density changes as a lithospheric plate moved away from an accretionary ridge. Isostatic adjustment under sedimentary load plays only a subsidiary role in "geosynclinal" subsidence (Hsü 1965).

Likewise, the Rhine Graben or the Great Basin could be regarded as a modern analogue of taphrogeosynclines. Coastal and offshore basins of Southern California apparently portray epi-eugeosynclines. Trenches and island arcs appear to represent embryonically folded orthogeosynclines. Thinly veneered ocean floors certainly qualify as leptogeosynclines. Those actualistic models provide vivid and concrete illustrations, and one begins to wonder indeed if the geosyncline concept has outlived its usefulness.

Now I would ask my readers to tolerate one last digression. Last summer as I wandered about the Mediterranean, I became an avid reader of post-Homeric interpretations of the *Odysseus*. Homer spoke romantically of the Lotus Eater's Land, the Scylla and Charybdis, the Rock of Calypso, etc. He made good poetry but left behind a poor textbook on Mediterranean geography. Many of us had read the saga repeatedly without having any idea where Ulysses went. Attempts to speculate on Homerian geography in the *Odysseus* have not been as spectacularly successful as Schliemann's reading of the *Iliad*. Nevertheless, references to actual places in the Mediterranean gave a new dimension to the Odysseus epic. A comparison of the Strait of Messina with the Scylla and Charybdis described by Homer provides a calibration of his poetic license.

This parable serves to summarize my impression of what we have learned during the last seventy years on the origin and nature of geosynclines. I see a tendency for ancient sediments to be more and more interpreted on an actualistic basis. References to modern geography do indeed make sense, and progress will come when we no longer need to hide our ignorance in myths and semantic traps. We now have sufficient means to sail across the Mediterranean; a trip through the Strait of Messina would provide us with a far more accurate, though perhaps less romantic, portrayal of its geographical settings than we could have learned by reading Homer's description of Ulysses's venture through the Scylla and Charybdis. On the other hand, we should not lose sight of the fact that a large body of geological data has been presented within the framework of the geosynclinal nomenclature. Geosynclinal terminology, when accurately defined, has served and will continue to serve as useful abbreviations for packages of rocks in folded mountain systems. I do envision, though, that future development will emphasize the descriptive and more factual aspects of geosynclinal rock suites and that there would be less temptation for us to indulge in speculative or dogmatic pronouncements that are unrelated to actualistic settings.

Looking back one might complain that we would have had no need to guess or to argue whether the Rock of Calypso is Malta or Pantellaria if Homer had given us more accurate descriptions and had been less extravagant with his poetic transgressions. But, then, there would have been no need for post-Homeric interpreters if Homer had not written the *Odysseus.*

REFERENCES

Arbenz, P. 1919. Probleme der Sedimentation und ihre Beziehungen zur Gebirgsbildung in den Alpen. *Naturf. Gesell. Zürich, Vierteljahresschrift, Jahrg. 64,* 246–75.

Argand, E. 1916. Sur l'Arc des Alpes Occidentales. *Eclogae Geol. Helvetiae 14,* 145–91.

———. 1922. La Tectonique de l'Asie. *XIIIe Congr. géol. intern., Bruxelles, Compt. Rend.,* pp. 171–372.

Aubouin, J. 1965. *Geosyncline.* Amsterdam: Elsevier Co., 335 pp.

Bader, R. G., et al. 1970. *Initial Reports of the Deep Sea Drilling Project, 4,* Washington: U.S. Government Printing Office, 753 pp.

Bailey, E. B. 1930. New light on sedimentation and tectonics, *Geol. Mag. 67,* 86–88.

———. 1936. Sedimentation in relation to tectonics. *Geol. Soc. America Bull. 47,* 1713–26.

Bailey, E. B., & McCallien, W. J. 1960. Some aspects of the Steinmann trinity, mainly chemical. *Geol. Soc. London Quart. Jour. 116,* 365–95.

Bailey, E. H., Irwin, W. P., & Jones, D. L. 1964. Franciscan and related rocks and their significance in the geology of western California. *California Div. Mines Bull. 183,* 177 pp.

Bailey, E. H., & Blake, M. C. 1969. Late Mesozoic sedimentation and deformation in western California. *Geotektonika 3,* 17–30; *4,* 24–34.

Barrell, J. 1914. The strength of the earth's crust. *Jour. Geology 22,*

Becke, F. 1903. Die Eruptivgebiete der böhmischen Mittelgebirge und der amerikanischen Anden. Atlantische und pazifische Sippe der Eruptivgesteine. *Tscherm. Min. Petr. Mitt. 22,* 209 pp.

Beloussov, V. V. 1962. *Basic problems in geotectonics.* New York: McGraw-Hill, 820 pp.

Beloussov, V. V., & Rudich, Y. M. 1961. Role of island arcs in the development of earth's structure. *Intern. Geology Rev. 3*, 557-74.

Bertrand, M. 1897. Structure des Alpes françaises et récurrence de certain faciès sédimentaires. *VIe Congr. géol. intern., Zürich, Compt. Rend.*, p. 163-77.

Brinkmann, R. 1933. Ueber Kreuzschichtung im deutschen Buntsandstein-becken. *Nachr. Ges. Wiss. Göttingen, Math.-physik. Kl. 1933*, pp. 1-2.

Bucher, W. H. 1933. *The deformation of the earth's crust.* Princeton: Princeton Univ. Press, 518 pp.

Cann, J. R. 1968. Spilites from the Carlberg Ridge, Indian Ocean. *Jour. Petrology 9*, 1-19.

Carey, S. W. 1958. The tectonic approach to continental drift. In *Continental Drift—A symposium*, Carey, S. W. *Convener.* Hobart, Australia: Univ. Tasmania, pp. 177-355.

Cloos, H. 1938. Primäre Richtungen in Sedimenten der rheinischen Geosyn-klinale. *Geol. Rundschau 29*, 357-67.

Dana, J. D. 1873. On some results of the earth's contraction from cooling, including a discussion of the origin of mountains and the nature of the earth's interior. *Am. Jour. Sci., ser. 3, 5*, pp. 423-43; *6*, pp. 6-14, 104-15, 161-72.

Dewey, H., & Flett, J. S. 1911. On some British pillow-lavas and the rocks associated with them. *Geol. Mag., decade 5, 8*, pp. 202-9, 241-48.

Dickinson, W. R. 1970. Second Penrose conference: the new global tectonics. *Geotimes 15, no. 4*, 18-22.

Dickinson, W. R., & Hatherton, T. 1967. Andesitic volcanism and seismicity around the Pacific. *Science 157*, 801-3.

Dietz, R. S. 1963. An actualistic concept of geosynclines and mountain-building. *Jour. Geology 71*, 314-43.

Dutton, C. E. 1889. On some of the greater problems of physical geology. *Phil. Soc. Washington Bull. 11*, 51-64.

Engel, A. E. J., & Engel, C. G. 1964. Igneous rocks of the East Pacific Rise. *Science 146*, 477-85.

Ewing, et al., 1970. *Initial Reports of the Deep Sea Drilling Project*, vol. 1. Washington: U.S. Government Printing Office, 672 pp.

Fischer, A. G. 1964. The Lofer cyclothems of the Alpine Triassic. *Kansas Geol. Survey Bull. 169*, 107-49.

Fisk, H. N. 1952. Sedimentation and orogeny with particular reference to the Gulf Coast geosyncline (abstract). *Geol. Soc. America Bull. 63*, 1328.

Gansser, A. 1964. *Geology of the Himalayas.* London: Wiley, 289 pp.

Gilbert, G. K. 1890. Lake Bonneville. *U.S. Geol. Survey, Monograph 1*, 438 pp.

Gilluly, J. 1935. Keratophyres of eastern Oregon and the spilite problem. *Am. Jour. Sci. 229*, 225-52; 336-52.

_____. 1949. Distribution of mountain-building in geologic time. *Geol. Soc. America Bull. 60*, 561-90.

_____. 1955. Geologic contrast between continents and ocean basins. *Geol. Soc. America, Special Paper 62*, 7–18.

Glaessner, M. F., & Teichert, C. 1947. Geosynclines: a fundamental concept in geology. *Am. Jour. Sci. 245*, 465–82; 571–91.

Griggs, D. T. 1939. A theory of mountain building: *Am. Jour. Sci. 237*, 611–50.

Hall, J. 1859. Paleontology. *Geological Survey of New York 3, pt. 1*, 66–96.

Hamilton, W. 1969. The volcanic central Andes—a modern model for the Cretaceous batholiths and tectonics of western North America. *Oregon Dept. Geol. & Mineral Industries, Bull. 65*, 175–84.

Harker, A. 1896. The natural history of igneous rocks: I. Their geographical and chronological distribution. *Science Progress, 6*,

_____. 1909. *The natural history of igneous rocks*. London: McMillan.

Haug, E. 1900. Les géosynclinaux et les aires continentales. *Bull. Soc. Géol. France, 3e ser., 28*, 617–711.

Heezen, B. C. 1968. The Atlantic continental margin. *Univ. Missouri Rolla Journal 1*, 5–25.

Heiskanen, W. A., & Vening Meinesz, F. A. 1958. *The earth and its gravity field*. New York: McGraw–Hill Co., 470 pp.

Hess, H. H. 1951. Comments on mountain building. *Trans. Am. Geophys. Union 32*, 528–31.

Hsü, K. J. 1958. Isostasy and a theory for the origin of geosynclines. *Am. Jour. Sci. 256*, 305–27.

_____. 1960. Texture and mineralogy of the Recent sands of the Gulf Coast. *Jour. Sedimentary Petrology 30*, 380–403.

_____. 1965. Isostasy, crustal thinning, mantle changes, and the disappearance of ancient land masses. *Am. Jour. Sci. 263*, 97–109.

_____. 1967. Mesozoic geology of the California coast ranges—a new working hypothesis, in *Etages tectoniques*, Schaer, J. (ed.) Neuchâtel: Baconnière, pp. 279–96.

_____. 1968. Principles of mélanges and their bearing on the Franciscan–Knoxville paradox. *Geol. Soc. America Bull. 79*, 1063–74.

_____. 1971. Franciscan mélanges as a model for eugeosynclinal sedimentation and underthrusting tectonics. *Jour. Geophys. Research 76*, 1162–70.

Hsü, K. J., & Ryan, W. B. F. 1971. Implications concerning ocean-floor genesis and destruction from deep-sea drilling in the Mediterranean Sea and eastern Atlantic Ocean. *I.U.G.G., 15th General Assembly, Moscow, Abstracts*, 4–30.

Jeffreys, H. 1924. *The earth*. Cambridge: University Press, 277 pp.

Jones, O. T. 1938. On the evolution of a geosyncline. *Geol. Soc. London, Quart. Jour. 94*, lx–cx.

Kay, M. 1951. North American geosynclines. *Geol. Soc. America Mem. 48*, 143 pp.

Kennedy, G. C. 1959. The origin of continents, mountain ranges, and ocean basins. *Am. Scientist 47*, 491–504.

Kennedy, W. Q., & Anderson, E. M. 1938. Crustal layers and the origin of magmas. *Bull. volcanologique, ser. II 3*, 24–82.

Knight, S. H. 1929. The Fountain and Casper formations of the Laramie Basin. *Univ. Wyoming Publ. Sci., Geol.,* 1.

Knopf, A. 1948. The geosynclinal theory. *Geol. Soc. America Bull. 59,* 649–70.

Kossmat, F. 1921. Die mediterranen Kettengebirge in ihrer Beziehung zum Gleichgewichtszustande der Erdrinde. *Sächsische Akad. Wiss. Math.-Phys. Kl. 38,* no. 2, 46–48.

Krumbein, W. C., Sloss, L. L., & Dappeles, E. C. 1949. Sedimentary tectonics and sedimentary environments. *Am. Assoc. Petroleum Geologists Bull. 33,* 1859–91.

Krynine, P. D. 1941. Differentiation of sediments during the life history of a landmass (abstract). *Geol. Soc. America Bull. 52,* 1915.

Kuenen, P. H., & Migliorini, C. I. 1950. Turbidity currents as a cause of graded bedding. *Jour. Geology 58,* 91–127.

Laubscher, H. 1969. Mountain building. *Tectonophysics 7,* 551–63.

Lawson, A. C. 1942. Mississippi delta–a study in isostasy. *Geol. Soc. America Bull. 47,* 1691–1712.

McKee, E. D., & Weir, G. W. 1953. Terminology for stratification and cross-stratification. *Geol. Soc. America Bull. 64,* 381–90.

Melson, W. G., & Van Andel, T. H. 1966. Metamorphism in the Mid-Atlantic Ridge, 22 N. latitude. *Marine Geology 4,* 165–86.

Migliorini, C. 1943. Sul modo di formazione dei complessi tipo macigno. *Boll. d. Soc. Geol. It. 62,* XLVIII.

Moore, J. 1965. Petrology of deep-sea basalt near Hawaii. *Am. Jour. Sci. 263,* 40–52.

Moores, E. 1969. Petrology and structure of the Vourinos Ophiolitic Complex of northern Greece. *Geol. Soc. America Special Paper 118,* 74 pp.

Natland, M. L. 1933. The temperature- and depth-distribution of some Recent and fossil foraminifera in the Southern California region. *Scripps Inst. Oceanography Bull., tech. ser. 3,* 225–30.

Natland, M. L., & Kuenen, P. H. 1951. Sedimentary history of the Ventura Basin, California, and the action of turbidity currents. *Soc. Econ. Paleontologists and Mineralogists, Special Publ. 2,* 76–107.

Neumayr, M. 1875. *Erdgeschichte, 1,* Leipzig: Bibliographisches Inst., p. 364.

Neumayr, M., & Suess, F. E. 1920. *Erdgeschichte, 1,* 3rd edition. Leipzig: Bibliographisches Inst., 543 pp.

Pakiser, L. C. 1963. Structure of the crust and upper mantle in the western United States. *Jour. Geophys. Research 68,* 5747–56.

Pettijohn, F. J. 1943. Archean Sedimentation. *Geol. Soc. America Bull. 54,* 925–72.

_____. 1950. Turbidity currents and graywackes–A discussion. *Jour. Geology 58,* footnote 3.

_____. 1957. *Sedimentary rocks.* New York: Harper, 718 pp.

Potter, P. E., & Pettijohn, F. J. 1963. *Paleocurrents and basin analysis.* Berlin: Springer, 296 pp.

Quon, S. H., & Ehler, E. G. 1963. Rocks of northern part of Mid-Atlantic Ridge. *Geol. Soc. America Bull. 74,* 1–8.

Reiche, P. 1938. An analysis of cross-lamination: the Coconino sandstone. *Jour. Geology 46*, 905–32.

Russell, R. D. 1940. Quaternary history of Louisiana. *Geol. Soc. America Bull. 51*, 1199–1234.

Ryan, W. B. F., et al. 1970. Deep-sea drilling project: Leg 13. *Geotimes 15*, no. 10, 12–15.

Scheumann, K. H. 1932. Ueber die petrogenetische Abteilung des roten Erzgebirgsgneisses. *Min. Petrog. Mitt. 42*, 423–26.

Schuchert, C. 1923. Sites and natures of the North-American geosynclines. *Geol. Soc. America Bull. 34*, 151–260.

Schuiling, R. D. 1969. A geothermal model of oceanization. *Verh. Kon. Ned. Geol. Mijnb. Gen. 26*, 143–48.

Sheridan, R. E., Drake, C. L., Nafe, J. E., & Hennion, J. 1966. Seismic refraction study of continental margin east of Florida. *Am. Assoc. Petroleum Geologists Bull. 50*, 1972–91.

Staub, R. 1928. *Der Bewegungsmechanismus der Erde*. Berlin: Borntraeger, 270 pp.

Steinmann, G. 1905. Die geologische Bedeutung der Tiefseeabsätze und der ophiolithischen Massengesteine. *Ber. naturf. Ges. Freiburg 16*, 44–65.

――――. 1926. Die ophiolithischen Zonen in dem mediterranen Kettengebirge. *14th Intern. Geol. Congress, Madrid, Compt. Rend. 2*, 637–67.

Stille, H. 1924. *Grundfragen der vergleichenden Tektonik*. Berlin: Borntraeger, 443 pp.

――――. 1936. The present tectonic state of the Earth. *Am. Assoc. Petroleum Geologists Bull. 20*, 848–80.

――――. 1940. *Einführung in den Bau Amerikas*. Berlin: Borntraeger, 717 pp.

Stokes, W. L. 1947. Primary sedimentary trend indicators applied to ore-finding in the Carrizo Mountains, Arizona and New Mexico. *U.S. Atomic Energy Comm., RE-3043, part 1.*

Suess, E. 1875. *Die Entstehung der Alpen*. Wien: W. Braumüller, 168 pp.

――――. 1909. *Das Antlitz der Erde, 3*, pt. 2. Leipzig: G. Freytag, 789 pp.

Tatel, H. E., & Tuve, M. A. 1955. Seismic exploration of a continental crust. *Geol. Soc. America, Special Paper 62*, 35–50.

Tilley, C. E. 1950. Some aspects of magmatic evolution. *Geol. Soc. London Quart. Jour. 106*, 37–61.

Trümpy, R. 1960. Paleotectonic evolution of the central and western Alps. *Geol. Soc. America Bull. 71*, 843–908.

――――. 1970. Stratigraphy and mountain-building. *Geol. Soc. London Quart. Jour.* (in press).

Tyrrell, G. W. 1933. Greenstones and graywackes: *Reunion Internationale pour l'Etude du Précambrien et des Vieilles Chaines de Montagnes en Finlande*, 1931, *Compte Rendu*, 24–26.

Van Bemmelen, R. W. 1958. Stromingstelsels in de silicaatmantel. *Geol. Mijnb., N.S. 20*, 1–17.

Van der Gracht, W. A. J. M. van Watershoot. 1931. Permo-carboniferous orogeny in the south-central United States. *Am. Assoc. Petrol. Geologists Bull. 15*, 991–1057.

Van der Gracht, W. A. J. M. van Watershoot, et al. 1928. Theory of continental drift. *Am. Assoc. Petrol. Geologists*, Tulsa, 240 pp.

Von Bubnoff, S. 1937. Gebirgsgrund und Grundgebirge. *Naturwissenschaften* 25, 577–88, 593–98.

Walther, J. 1897. Ueber Lebensweise fossiler Meeresthiere. *Zeitschr. Deutsch. Geol. Ges. 49*, 209–73.

Wegener, A. 1912. Die Entstehung der Kontinente. *Dr. A. Petermann's Geogr. Mitt.*, 1912, April Heft, no. 4, 20 pp.

———. 1922. *Die Entstehung der Kontinente und Ozeane.* Braunschweig: F. Vieweg & Sohn, 144 pp.

Wells, F. C. 1949. Ensimatic and ensialic geosynclines (abstract). *Geol. Soc. America Bull. 60*, 1927.

Winterer, E. L., & Durham, D. L. 1962. Geology of southeastern Ventura Basin, Los Angeles County, California. *U.S. Geol. Survey, Prof. Paper 334-H*, pp. 276–366.

Woodford, A. O. 1960. Bedrock patterns and strike slip faulting in southwestern California. *Am. Jour. Sci. 285 A*, 440–47.

Chapter Four

Concepts of Appalachian Basin Sedimentation

EARLE F. McBRIDE

INTRODUCTION

A sedimentologic interpretation during any period in history reflects the technological state of the art in general and the specific training, experience, and wisdom of the individual. This essay examines the changing interpretations of the origin of various sedimentary rocks in the Appalachian Basin; it will show how both interpretations and philosophies of study have changed during the past seventy years, largely in response to new developments in the understanding of Recent sediments. The essay will also emphasize the need to integrate sedimentologic and tectonic concepts to help unravel global tectonic history.

The birthplace of geology in North America is in the eastern part of the United States, largely within the broad limits of the Appalachian geosyncline. The unmetamorphosed filling of the geosyncline includes sedimentary rocks of Late Precambrian and Paleozoic ages bordered by Piedmont crystalline rocks along the eastern edge, by Mesozoic cover along the southern edge, and by an arbitrarily placed western margin along the Nashville–Cincinnati arches. Geosynclinal rocks extend north through New England into the Maritime provinces of Canada, but only those rocks south of the Adirondack massif in New York are considered here. An excellent stratigraphic-sedimentologic summary of the Appalachian Basin was presented recently by Colton (1970).

The Appalachian Basin and adjoining areas were studied initially by the early state geological surveys and academicians of nearby universities and museums, and later by workers of the United States Geological Survey. The accumulation of basic stratigraphic and paleontologic data led inevitably to concepts of the sedimentologic and structural history of the region. Some concepts were directed to local features, whereas others were more general and applicable to other regions as well. Because so many geologists received part of their field training in the Appalachian Basin, and because its rocks are so rich in natural resources, geologic data and attendant ideas have continued to accumulate at an ever increasing rate. After reviewing the published sedimentologic reports from a historical viewpoint I believe the following major points can be documented:

1) Early interpretations, which at first glance appear primitive by today's standards, include many astute conclusions.

2) Detailed sedimentological interpretations were not possible until a reasonable understanding of sedimentary processes developed.

3) Most sedimentologic studies are myopic and consider only one or two aspects, such as texture or structure. Most stratigraphic studies are hyperopic and consider only the broad questions of rock type, geometry, age, and regional correlation. Both approaches yield valuable conclusions, but what is needed is more syntheses utilizing both approaches.

4) Field work has provided the most productive data for first-order sedimentological interpretations.

5) The evolution of concepts, techniques, and the philosophy of Appalachian sedimentation reflects the "state of the art" of sedimentology in general and I cannot identify a concept that is unique to the Appalachian Basin.

6) Since the mid-1950's there has been an explosive increase in publications on the sedimentology of the Appalachians.

Literature Compilation

One approach to show the evolution of sedimentologic concepts in Appalachian Basin sedimentation is to examine changes in approaches and topics of interest of the published record. A search of the literature on articles written on rocks of the Appalachian Basin in the past seventy years whose first or second major topic concerns a sedimentologic theme yielded 135 contributions. This list ignores abstracts, theses and dissertations, and includes articles of publications in the library of the University of Texas at Austin. I included only articles that recorded some qualitative or quantitative data other than simple megascopic descriptions of rocks. This biased approach omitted from my list some otherwise important contributions on concepts, such as articles by Grabau and Barrell, and omitted important stratigraphic works by many workers. Some of the data collected are shown in graphs that have five-year class intervals from 1900 to 1970 as the ordinate (Figs. 1–6). The total number of articles with sedimentologic themes (Fig. 1) shows about a *ten-fold* increase since 1955. The total number of pages written increases accordingly (Fig. 2), and the average length of pages per article has increased slightly and rather unevenly. Articles on the coarse terrigenous rocks (sandstone and conglomerate) have been more numerous than those on

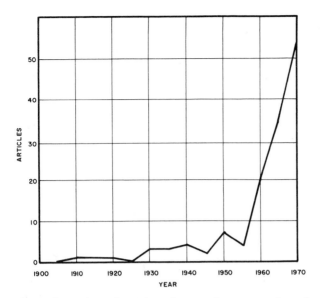

Fig. 1. Plot of number of articles whose main or secondary theme concerns some sedimentologic aspect of Appalachian Basin rocks versus dates since 1900. Articles (total of 135) are grouped into five-year class intervals. See text for additional data.

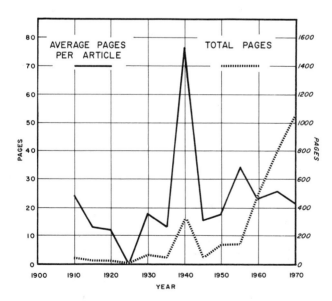

Fig. 2. Plot of (1) total number of pages and (2) average number of pages per article versus date for same articles plotted in Figure 1.

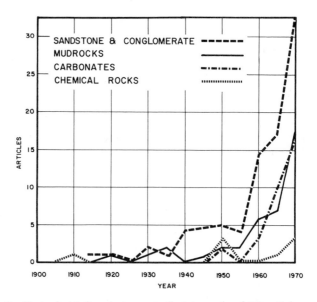

Fig. 3. Plot of chief rock type studied in each of 99 articles on Appalachian Basin rocks versus date.

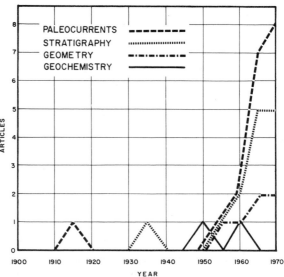

Figs. 4-6. Plot of theme of articles versus years for same data shown in Figure 1. Data includes themes (aspects or techniques of study) that are of first or secondary importance in each article.

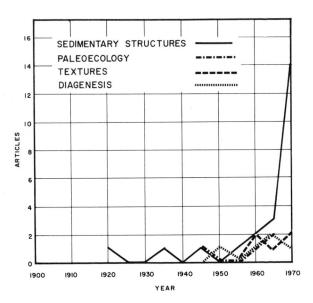

Fig. 5.

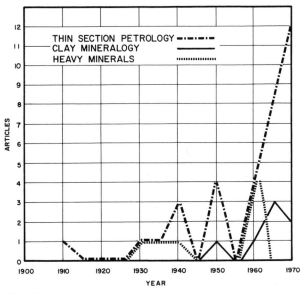

Fig. 6.

other rock types, although articles on carbonate rocks are increasing rapidly (Fig. 3). Figures 4–6 show the distribution of articles whose major sedimentologic topic is cited. Conspicuous trends include the major role that thin-section petrology has always played and the spectacular rise in paleocurrent studies beginning in 1955 and articles on sedimentary structures since 1965.

It is interesting to speculate where we are today on the logarithmically rising curve of knowledge; at the base of the rise? half way to the asymptote? or near its top?

Refined Interpretations of Environment and Processes of Deposition

Up to 1900 and even midway through the last seventy years most interpretations of conditions of deposition of Appalachian Basin rocks were made by geologists whose chief information was stratigraphic data or megascopic petrographic data. The major concern was whether the deposits were subaerial or subaqueous, and if subaqueous whether fresh, brackish, or marine. If marine, the problem was whether the water was quiet and clear or turbulent and muddy. The major environmental categories identified were warm, shallow, marine water; continental; or rarely, deltaic or eolian environments. Conclusions were based chiefly on stratigraphic data; on the presence of ripple marks, mud cracks, and the colors of rocks; on gross paleoecologic interpretations deduced from fossils; and on a generally often incorrect understanding of sedimentary processes.

Although there were some significant contributions to paleogeography and sediment provenance during these early years, the major breakthrough in interpreting Appalachian Basin rocks has come in the past fifteen years, from a better understanding of recent sediments and increased knowledge of the sedimentary processes operating in specific sedimentary environments. To illustrate this major breakthrough in the development of new concepts, I will compare the changing interpretations of three formations.

1. **Tuscarora–Shawungunk–Green Pond Formations (Silurian).** These formations comprise the remnants of a clastic wedge composed of mineralogically mature quartz-pebble conglomerate in the east, grading to medium sandstone to the west. The maximum thickness of the complex is in excess of 1,000 feet. Stratigraphic summaries of the units are given by Swartz and Swartz (1931), Yeakel (1962), and Meckel (1970). A major problem is to explain the origin of such a thick, mature quartz-sediment wedge that essentially lacks shale beds. Interpretations include:

Clarke and Ruedemann (1912, p. 105): a "tidal" origin.

Grabau (1913, p. 468): "the fact that fossils are absent . . . would indicate that these beds were deposited above the sea, along river floodplains."

Schuchert (1916, p. 535): "deposits are of the shallow sea and near the shore—great shallow-water flats of moving sands."

Yeakel (1962, p. 1534): deposition was by "lateral migrating (meandering?)" streams to "build up a thick succession of channel-deposited sands in the form of essentially lenticular sheets. Rates of subsidence and reworking of material during lateral migration seems to have operated collectively in removing most fine-grained flood-plain deposits."

Smith (1970, p. 2993): "These formations display downslope trends similar to those of the South Platte and Platte Rivers, and, combined with paleocurrent, grain-size distribution, and other data suggest that the coarse eastern facies (Green Pond, Shawungunk) represent proximal braided stream deposits with longitudinal bars that grade westward and northwestward into finer-grained distal braided stream sediments (Tuscarora) characterized by transverse bars."

2. Martinsburg Formation (Ordovician). This flysch unit crops out in the Great Valley Province from New York to Virginia and is locally nearly 10,000 feet thick. Gray shale and interbedded graded graywacke and gray shale characterize the formation (Behre 1927; McBride 1962). The main problem was to explain the mechanism of sandstone deposition and interpret the depth of deposition. Interpretations include:

Lesley (1892, p. 565): "it is impossible to suppose 5,000–6,000 feet of mud to have been deposited in any but a deep basin." He elsewhere (p. 527) quotes the opinions of colleagues who interpreted these rocks in New York to be shallow-water deposits.

Behre (1927, p. 84): "That [the Martinsburg sea] . . . had a fairly constant depth, so that sediment was distributed evenly over most of its bottom, is shown by the uniformity of the sequence, bed for bed. . . . Shallow-water is indicated by the ripple marks, cross-bedding, and rapid alternation of the various types of sediments."

Van Houten (1954), Pettijohn (1957, p. 615), and McBride (1962) interpreted the sandstones to be turbidites that were deposited in fairly deep water. I used flute and groove cast data (McBride 1962, Fig. 32) to infer that turbidity currents flowed laterally into a linear trough-like basin and then changed course to flow longitudinally down the axis. Northeasterly flowing currents and southwesterly flowing currents converged at the present New Jersey–Pennsylvania boundary, the presumed deepest part of the basin.

Klein (1967, p. 378) suggested an alternate hypothesis for the sandstones that show longitudinal transport. He suggested that sediment transport was by geostrophic, contour-following ocean-bottom currents rather than by turbidity currents, and that the two current systems converged near the New Jersey–Pennsylvania boundary where the lighter current overrode the denser current.

3. Manlius Formation (Devonian, New York). The stratigraphy and paleontology of the Manlius Formation (less than 200 feet thick) and associated Lower Devonian carbonate rocks have been the subject of many studies over that past 100 years. For these limestone and dolomite rocks, and for carbonate rocks in the Appalachian Basin in general, a shallow-water marine origin has been almost an *a priori* interpretation. Rickard (1962, pp. 94–96) used sedimentary structures, lithology, and fossils as the bases for interpreting a shallow-water lagoonal environment for the bulk of the formation and a more agitated environment for stromotoparoid biostromes. Laporte (1967) used studies of recent carbonate accumulations in Florida and the Bahamas to identify supratidal, intertidal, and very shallow subtidal deposits in the Manlius. This surprisingly precise interpretation of tidal zones no more than several feet thick at any one time is dependent on searching comparison of recent and ancient sediments; it has opened the door to a whole new level of understanding of depositional environment.

Different Approaches to Interpreting Provenance

One of the major goals in the sedimentological study of terrigenous clastic rocks is to determine the distance and direction the detritus was transported, and deduce the climate and relief of the source area (Pettijohn 1957, p. 498). Although the concepts or approaches to studies of provenance were not invented by workers in the Appalachian Basin, their papers document the evolution of the ideas and techniques that are in common use today. A chronologic summary of major contributions dealing with the concept of the old landmass of Appalachia as the main source of terrigenous debris in the Appalachian Basin is treated in a later part of this essay. The point I wish to make here is that several different disciplines have been used to interpret source area and provencance.

1. Stratigraphic approach. Basin data on regional variations in thickness of formations and lateral changes in coarseness of detritus permitted early workers to infer the direction of sediment transport. Megascopic study of clasts, chiefly gravel-sized, gave information on source-area composition. Barrell used red beds as a criterion of climate in both the source area and environment of deposition. Barrell (1912, 1914) maintained that red beds formed subaerially in a semi-arid climate with seasonal rainfall. Semi-arid conditions developed on the leeward (western) flank of mountainous Appalachia in the rain shadow, according to the interpretation of Grabau (1913, p. 442).

2. Thin-section approach. Krynine's (1940) study of the Third Bradford Sand is a classic model of the sorts of interpretations of environment of deposition, provenance, and transport that can come from analysis of thin-sections. Mineralogy and texture were used to infer the rock composition, climate (rainfall, humidity), relief, and type of weathering of source rocks, as well as rates of erosion and sediment transport. This detailed approach requires either stratigraphic or paleocurrent data to determine the direction to the source area.

3. Paleocurrent approach. Although oriented fossils (Ruedemann 1897, pp. 367–91) and the orientation of ripple marks (Hyde 1911) had been used earlier to determine paleocurrents in the Appalachian basin, the practice of systematically recording directional structures to determine paleocurrent patterns of sediment transport in this basin has been championed by F. J. Pettijohn and implemented beginning in the 1950's by his students. The philosophy of the paleocurrent approach was summarized by Pettijohn (1962), and Potter and Pettijohn (1963), and the measurement of directional structures is now a conventional technique. I will not attempt to document the importance of this study approach to Appalachian Basin history, because reviews of this nature have recently appeared (Meckel 1970; McIver 1970) and new data is accruing rapidly (Adams 1970; Brown 1970; Hunter 1970; Schwab 1970; Whisonant 1970). Although stratigraphic data alone (e.g., Colton 1970) can tell the direction of terrigenous source areas, paleocurrent studies integrated with detailed mineralogical studies yield more detailed interpretations (e.g., Pelletier 1958; Yeakel 1962). But, independent of source direction, paleocurrent data yields information on patterns of currents that are clues to processes of deposition (Adams 1970), bathymetry of the marine basin (McBride 1962; Klein 1967), and downstream increase in rate of meandering in fluvial deposits (Pelletier 1958, p. 1062).

CONCEPT OF DELTAIC SEDIMENTATION

General

The history of the concept of deltaic deposition in the Appalachian Basin documents the importance of understanding sedimentary processes as a means of interpreting rocks, the necessity of a regional knowledge of facies within a stratigraphic package for making interpretations, and the inadequacy of myopic sedimentologic studies to provide convincing interpretations of regional scope.

Prior to Joseph Barrell's writings on deltaic sedimentation (1908–14), deltas were not considered important in the sedimentary history of the Appalachian Basin. The Atlantic coast today lacks deltas but has many estuaries, so it is not surprising that estuarine environments were preferred to deltaic environments when interpreting ancient deposits. The deposits of coastal swamps and marshes were inferred in some reconstructions, but not interpreted as being associated with deltas. Barrell made a strong argument for a deltaic origin of the Mauch Chunk (Mississippian) red beds (1908) and for the better known Catskill (Devonian) red beds (1913, 1914) and set forth criteria for recognizing ancient deltaic deposits (1912). His definition of a delta (Barrell 1912, p. 38) as "a deposit partly subaerial built by a river into or against a body of permanent water" is still quoted as a workable definition. Barrell said a delta was characterized by its major stratigraphic relations of topset, foreset, and bottomset beds; by overlapping of these facies resulting from progradation; and by a cyclic habit. Deltaic deposits were recognized by sedimentary structures, color, and fossils that established subaerial deposits adjacent to submarine deposits.

During the next fifty years deltas received little consideration until the appearance of the classic paper on the Bedford–Berea deltas (Mississippian) by Pepper et al. in 1954. Largely from subsurface data these authors established a deltaic interpretation on (1) a facies pattern of red beds surrounded by marine deposits; (2) upward convexity of the Bedford Formation that resembles the active Mississippi delta, (3) flanking offshore bars, and (4) a deltoid shape of the deposits.

Increased precision in the interpretation of Appalachian deltaic deposits has come only in the last decade, when detailed comparisons between ancient and modern deltaic facies were attempted. One example of this increased precision is the coal-bearing Carboniferous rocks of the greater Dunkard Basin. Williams and Ferm (1964), Ferm and Williams (1965), Ferm and Cavaroc (1968), and Ferm (1970) use the Holocene Mississippi alluvial and deltaic plains as a model for interpretating these Carboniferous rocks, whereas Donaldson (1969) uses the modern Colorado River delta in Texas. Donaldson, *ibid.* makes a close comparison between ancient (Conemaugh and Monongahela groups) and modern deltaic facies of a rapidly prograding shallow-water delta (Colorado River delta model) in contrast to facies of a rapidly prograding deep-water delta like the Belize lobe of the modern Mississippi. The ability to relate rock geometry, sedimentary structures and textures, and fossils to specific environments and processes provides a highly convincing argument for deltaic sedimentation.

Another study that uses the Mississippi delta as its model is the analysis by Wanless et al. (1970) based largely on subsurface data of Late Paleozoic deposits extending from Pennsylvania to Oklahoma.

The increasing emphasis on deltaic environments is climaxed in Colton's (1970, p. 8) statement that most sediments in the eastern part of the Appalachian Basin accumulated in deltaic environments.

EVOLUTION OF THE DELTA CONCEPT

In the early part of this century deltaic deposits were not believed to be a major component of the Appalachian Basin. I believe this bias came from the absence of well-documented recent examples and too little understanding of sedimentary processes. Proponents of deltaic sedimentation were in the minority. A deltaic origin of Late Paleozoic coal was favored by some workers, but there were formidable opponents, such as J. D. Dana (1856, p. 331), who wrote:

The American rocks throw much light on the origin of coal. Professor Henry D. Rogers, in an able paper on the American coal-fields, has well shown that the condition of a delta or estuary for the growth of the coal-plants, admitted even now by some eminent geologists, is out of the question, unless the whole continent may be so called; for a large part of its surface was covered with the vegetation. Deltas exist where there are large rivers; and such rivers accumulate and flow where there are mountains. How, then, could there have been rivers, or true deltas of much size, in the Coal Period, before the Rocky Mountains or Appalachians were raised? It takes the Andes to make an Amazon.

Red-bed formations that Barrell (1908, 1913, 1914) interpreted as predominantly deltaic had a history of varied interpretations. Lesley (1895, pp. 1806-07) believed the Mauch Chunk (Mississippian) was deposited "on a broad shore-bordered lowland near the sealevel, and in regions of its wide extent occupied by marshes, pools, and lagoons on which the first true coal vegetation began to grow." For the same rocks Willis (1902, p. 69) dismissed a deltaic interpretation because: "As the conditions became favorable for them, strong currents circulated. . . . They checked the development of deltas by sweeping the river mud away to greater depths."

In 1912-14 deltaic sedimentation got a large boost from Barrell (162 published pages) and Grabau (130 published pages). In 1912 Barrell treated the concept of deltaic sedimentation in general, but was concerned chiefly with distinguishing between subaerial and marine deposits and those strand-line deposits between them. His criteria of deltaic strand-line deposits can apply to tidal-flat deposits of non-deltaic areas equally well. Although Barrell believed topset, foreset, and bottomset beds of some ancient deltas could be identified, he warned (1912, p. 386-90) that the "textbook" delta model of Gilbert (1885) was given more attention than it deserved. He noted that Gilbert-type deltas (1) were small lake deltas, (2) were composed of coarse sediment, (3) were overweighted in significance compared to other types of deltas because of Pleistocene glaciation. He also emphasized that foreset beds could have slopes less than $2°$ or be totally absent in Holocene deltas.

Barrell's (1913, 1914) concept of deltaic origin of the Catskill sequence is nicely summarized by a critic of his (Mencher 1939, p. 1765) who wrote:

> The Catskill beds were deposited as the subaerial topset beds of a large delta forming along the northeastern part of the shallow Appalachian geosyncline. The Chemung and other marine beds were the subaqueous topset beds into which the land deposits graded. The source of most of the sediments was from the east but most particularly from the southeast from the western part of the old Appalachian landmass. The source was therefore about 200 miles from the easternmost part of the deposit. The Skunnemunk conglomerate, in its outlier 25 miles southeast

of the Catskill front, represented the piedmont gravel plain that continued to the base of the mountains. The materials were deposited by a series of short streams somewhat in the manner of a number of coalescing fans and not by one or two large rivers. The areal extent of the deposits was a good deal farther than the outcrops now show. The Atlantic faunal province was connected with the Appalachian faunal province by a northern rather than a southern one [as was usually inferred by other geologists at the time]. The climate was warm with seasonal rainfall. Comparable formations are the Siwalik of India and the more recent Indo-Gangetic plain deposits [which in a former paper Barrell (1906) had used as a type for his piedmont river deposits] and the Tertiary to recent alluvial deposits facing the Rocky Mountains.

Barrell based his conclusions on a study of the general field relationships of the formations, the character of the various rocks and their gradations, and the sedimentary structures present, such as cross-bedding, ripple marks, and mud cracking.

Mencher (1939) argued that the Catskill facies was largely alluvial plain deposits with some red shales being flood plain and deltaic subaerial topset deposits. He criticized Barrell for including fluvial deposits within his topset deltaic facies. Although this is a valid criticism, it is clear (Friedman & Johnson 1966, p. 177) that Barrell realized that topset deltaic rocks were in part fluvial in origin and that they graded laterally into fluvial deposits.

Grabau (1913) was even less precise in his use of "deltaic" because he repeatedly refers to deposits of dry deltas, normally called alluvial fans, and of wet deltas (conventional meaning).

Pepper et al. (1954) reconstructs an elaborate history of deltaic deposition of the "red" Bedford and Berea formations in Ohio during Mississippian time. Their series of colored paleogeographic maps show the southward progradation of the deltas and their subsequent decay. Other plates show the regional paleogeography of the central Appalachian Basin with deltas prograding westward into Pennsylvania and the Virginias. Although some of their criteria for a deltaic interpretation are not necessarily definitive (flanking offshore bars, upward convexity of main clastic bodies), their regional stratigraphic data and ability to map channel sandstones into what at that time appeared to be sheet-like bodies are convincing arguments.

In his detailed petrographic study of the Third Bradford Sand (Mississippian), Krynine argued for a deltaic origin on far shakier evidence. Although Krynine used published subsurface data to aid his interpretation, his own contribution was based on samples, cores, and thin sections chiefly from six wells. Krynine concluded (1940, p. 82):

> The deltaic character of the basin of deposition is inferred from many criteria. The abundance of fine-grained material indicates an abrupt pell-

mell deposition which may have been caused either by a closed basin (no outlet, everything is trapped) or by rapid flocculation of the colloidal and near-colloidal matter due to increased salinity. A closed basin does not need to be a structurally closed one, because an effective barrier may be set at the confluence of currents moving in different directions. Very probably both factors were operative in the Bradford delta.

The deltaic character is furthermore substantiated by an alternation of marine fossils and carbonized wood, by the changes in the physical character of the clay (loose particle vs. aggregates, a feature which may well have been due to periodic changes in salinity) and finally by cross-bedding, channeling intraformational conglomerates, and other features usually thought of as fluvial but which on account of their extremely small, almost microscopic, scale of occurrence suggest very slow-moving water indeed.

The dark colors of the sediments indicate deposition entirely below water level, in a reducing basin to which access of oxygen was denied. This points to the probability that such parts of the Bradford delta as we know were not formed under subaerial conditions.

Finally, the large-scale lensing and channeling of the Third Sand appears to correspond in many respects to the published descriptions of large deltas such as that of the Mississippi.

Although Krynine's study remains as a classic example of the detail in interpretation toward which sedimentologists strive, it also illustrates the need for a proper understanding of sedimentary processes and of regional facies relations in making environmental interpretations.

Post-Barrellian sedimentological interpretations of Catskill and equivalent marine facies are continuing; I found twenty contributions published since 1962. Whereas some writers tacitly accept of support a deltaic origin for parts of the sequence (e.g., Wolff 1965; Friedman & Johnson 1966; Sutton et al. 1970), others suggest that deltaic deposits are trivial or local. Allen and Friend (1968) present a coherent general environmental reconstruction of the Catskill–Chemung–Portage facies, but reject a deltaic settling. They conclude (pp. 58–59) that "rivers debouched into a complex marginal environment of tidal flats and subtidal shoals," seaward of which were barrier islands and an open shelf. A different nondeltaic interpretation for Catskill rocks in central Pennsylvania is presented by Walker (1971) and Walker and Harms (1971), who describe progradation of a shoreline that was supplied mud by long-shore currents from a distant delta in a low-wave, low-tide-range sea. In contrast, Friedman and Johnson (1966) interpret Catskill deltaic progradation, at least in New York, to have covered several hundred miles laterally and greater than 1,000 feet (post-compaction) vertically. These authors describe the sequence as a tectonic delta (containing a large sand to mud ratio derived from a nearby tectonic highland) and contrast it with the Mississippi and other Pleistocene and Holocene deltas. Clearly, data of modern processes and deposits are essential for sophisticated interpretations.

"THE DEVONIAN MOUNTAINS ARE GONE. . . . BUT IN REMNANTS OF FORMATIONS BORN OF DESTRUCTION WE MAY READ THE EPITAPH WHICH RECORDS THEIR GREATNESS" (Barrell, 1914, p. 253)

APPALACHIA AND RELATED SOURCE LANDS

Evolution of the Idea and Inferred Location of Appalachia

The sedimentary fill of linear geosynclines or of continental margins that become foldbelts yields evidence for the existence of landmasses marginal to the basins and away from the cratons. The concept of such a land, commonly called Appalachia, that supplied terrigenous debris to the Appalachian Basin throughout much or part of the Paleozoic Era grew as the stratigraphic data gathered in the 1800's permitted speculation on the position and nature of paleogeographic elements. Stratigraphic analysis was followed by sedimentologic analysis, from which tectonic hypotheses were proposed to "put it all together." The quest for Appalachia documents the interrelations of stratigraphy, sedimentology, and tectonics in the evolution of past and present ideas.

The concept of the permanency of ocean basins and continents was introduced by Dana (1846, pp. 354–55), who inferred the origin of these crustal topographic elements by shrinking of the cooling crust. Dana (1856*a*, p. 319) later expressed the idea that a "great reef or sand bank" situated at the cite of the present Appalachian Mountains served to separate a "vast continental lagoon" from the Atlantic Ocean. He inferred that sediment was brought to the Appalachian Basin by the southward-flowing Labrador current (Dana 1856*b*, p. 344) "which would have kept by the shore as now, along the eastern border of the Azoic land margin," and thence "over the Appalachian region, where the sandstones and shales were extensively accumulated; and therefore its aid in making these deposits can scarcely be doubted." Apparently the suggestion had been made that the sediments under discussion might have come from a landmass of continental character, because Dana states his opinion (1856, p. 331): "on the idea that the rocks of our continent have been supplied by sands and gravel from a continent now sunk. No facts prove that such a continent ever existed, and the whole system of progress . . . is opposed to it. . . . The existence of an Amazon [River] or any such Atlantic continent in Silurian, Devonian, or Carboniferous times is too wild an hypothesis for a moment's indulgence."

Thirty-four years later Dana (1890) used new information to formulate the idea that persistent positive elements originated during Archean time along the Atlantic and Pacific margins of the continent at the sites of the later-formed Appalachian and Cordilleran mountains. These ridges he called *protaxes*, and inferred them to be the source of much detritus. A granitic and gneissic composition of these Archean protaxes was implied. At the same time the role attributed to the Labrador current was abandoned because of the inference that at the end of Ordovician time the Green Mountains (his eastern protaxis) had emerged (Dana 1890, p. 43):

> Observe here what a blow the fact of this closed Northeast Bay gives the old theory—which I have held as well as others—that the coarse and fine sediment for Appalachian rock-making, during the Upper Silurian era and afterwards, came in, period after period, from the northeast, through Labrador currents. . . . It is the unavoidable conclusion that all the sedimentary beds of New York and the Alleghenies, through the Upper Silurian, Devonian, and Carboniferous eras, though so many thousands of feet thick, were made within the Interior sea out of material derived, so far as non-calcareous, from the wear of rocks about it, and that the tidal and other currents of the Interior sea distributed the material.

As chronicled by Kay (1951, p. 30), Wolcott (1891, pp. 363–69) reiterated the idea of the marginal ridge as a partial source of sediment, and in a footnote (p. 365) suggested the ridge might have extended to the Atlantic coastal plain, and might later have disappeared:

> It is not improbable that the area of the great coastal plain of the Atlantic slope was then an elevated portion of the continent and that much sediment deposited during Cambrian and later Paleozoic was washed from it into the seas immediately to the west. If this be true the source of much of the sediment of the Appalachian series of rocks is accounted for and the absence of the deposits of the eastern coast line is explained by the sinking of the coastal region during, or at the close of Paleozoic time.

The name "Appalachia" was coined by Williams (1897, p. 394), who applied it to a landmass of Devonian and later age whose shoreline extended from eastern New York to central Alabama, and whose "shores . . . were chiefly of Archean rocks."

In a treatise entitled "Paleozoic Appalachia or the History of Maryland during Paleozoic Time," Willis (1902) presented an artistic reconstruction of the limit, character, and history of Appalachia as viewed by him. His report includes two colored fold-out plates that show Appalachia to extend inland to include all of the Carolinas and most of Virginia, Maryland, and New Jersey; the eastern margin is beyond the modern Atlantic shoreline to a nebulous area labeled "Land or Atlantic Ocean." During Devonian time (Willis 1902, Plate VII) Appalachia is shown to include a southern "Province of the Southern Lowland" and a northern "Province of the Devonian High- land." He is specific in inferring Appalachia to have been a continent (p. 37), to have been composed of "siliceous crystalline rocks" (p. 62), and to have had highlands that "exhibited rounded forms of hill and valley, and thus may be contrasted with those ranges which present savage precipitous profiles" (p. 62). He also computed the volume of Middle and Late Devonian sedimen- tary rock in the Appalachian Basin and concluded that (p. 62): "If this mass with approximately the dimensions with which it was deposited in the sea, could be restored upon a sea-level plain of Appalachia, it would constitute a mountain range closely resembling in height, extent and mass the Sierra Nevada of California."

Barrell (1913, 1914), in his monumental treatise on the Catskill delta, used a similar approach and computed a volume of 63,000 cubic miles for Upper Devonian rocks alone (1914, p. 243). He inferred that the watershed of Appalachia was roughly 100 miles east of the present New Jersey shore. Barrell (1914, p. 245) also declared that proof of the source of Devonian sediments was clearly discernible from the "directions of delta growth, thick- ness [variation] of the Catskill beds, and the gradations in their texture." The demise of Appalachia, according to Barrell, was a combination of subsequent marine planation and the foundering of broken and fragmented parts of the landmass.

Schuchert popularized the idea that Appalachia was a major landmass through his paleogeographic maps in technical articles (1910, 1923) and text- books (e.g., Schuchert & Dunbar 1941), and "Appalachia" became a geologic household name long before it was identified with poverty-ridden terrain. He accepted Barrell's reconstruction of the size of Appalachia and showed con- cern for the eastern margin of the landmass (1923, p. 161):

> The question must now be asked, was there also a geosyncline to the east of the Appalachian one? Our answer is that there is nothing in what remains of the western part of Appalachia to show that such a trough ever existed in this borderland. Much further than this we can not go, but from Barrell's physiographic studies of Appalachia in Devonian times it is clear that if another geosyncline was present it must have lain upward of 200 miles east of the eastern shore of the Appalachian geosyncline. Keep- ing in mind the present depths of the Atlantic Ocean, however, we are disposed to believe that Appalachia was throughout a highland and of the nature of a geanticline.

Inasmuch as Barrell's estimate of the size of Appalachia was based on clastics (solids) alone, he surmised that Appalachia extended even farther than 250 miles from the present shore when dissolved constituents were added to the estimated restoration. He noted (1923, p. 161) "whatever the area and outward form of Appalachia, it was not an independent continent; rather was it an integral part of North America." Schuchert attributed the disappearance of Appalachia by (p. 162)"foundering into the depths of the Atlantic (Poseidon) seemingly as early as the Jurassic period."

A different possibility on the fate of Appalachia was eloquently expressed by Willard (1939, pp. 362-63) as the seeds of the concept of continental drift reached the United States:

> Along the eastern margin of Palamerica there extended a land mass. Its western shore approximated, during much of the era, that part of the continent called today the Piedmont. How far east the land may have reached is problematical. Estimates vary from only a few hundred miles to a vast land, a land now largely incorporated in the continent of Europe. Such a grotesque-sounding statement is based on the hypothesis that North America and Europe were once united as a single landmass which split asunder, the respective fragments drifting apart, leaving between a wide, water-filled rift, the Atlantic Ocean.

A totally different paleogeographic concept was being evolved by Kay from stratigraphic work spanning several years and summarized in his classic memoir (Kay 1951). His thesis is expressed in his introduction (pp. 4-5):

> The Paleozoic paleogeography of North America has been interpreted in the past on the theory that the continent was margined from the beginning of Paleozoic time by great lands of ancient crystalline rocks, Cascadia, Appalachia, and Llanoria on west, east, and south that persisted long as dominant sources of detritus and ultimately foundered beneath seas advancing from the bordering oceans. The writer comes to the contrary conclusion that the borders of North America are dominated by geosynclines, deposited beneath paleogeographic troughs adjoining and including linear tectonic and volcanic islands, analogous to the present island arcs (Pl. 1). This is the theory of marginal volcanic troughs and island arcs (Kay, 1944). The volcanic rock-bearing geosynclines lying beyond the miogeosynclines in the orthogeosynclinal belts, designated eugeosynclines, developed through long spans of time, were severely deformed and intruded, and superseded by other sorts of geosynclines, some of which are still forming.

Kay labeled the craton-ward (miogeosynclinal) trough the Champlain belt and the outer (eugeosynclinal) trough the Magog belt. Both troughs were inferred to receive their main source of detritus from volcanic sources situated within the geosyncline and also from "tectonic lands," island tracts of deformed and uplifted sediments originally deposited largely within the eugeosyncline. Although abandoning the concept of a continental mass for Appalachia, Kay (1951, p. 56) was willing to apply the term to the "tectonic land" that supplied the detritus that comprises the Late Devonian-Early Mississippian clastic wedge in the central Appalachians. However, the major role assigned to volcanism by Kay has been questioned by subsequent petrographic work in the same area (central but not northern Appalachians). Pettijohn (1970, p. 2) noted that seven major petrographic studies of the clastic rocks derived from Appalachia show a near absence of volcanic detritus; hence, the Kayian model has been viewed cautiously by many sedimentologists.

Concept of Composition of Appalachia

Williams's (1897, p. 394) definition of Appalachia as "chiefly Archean rocks" apparently implied that the mass was largely granite and gneiss. This concept persisted (Willard 1939, p. 362) until thin-section petrography of Appalachian Basin rocks was done. Mencher (1939) was one of the first to use this approach when he attacked Barrell's (1913, 1914) conclusion that the Catskill deltaic deposits came from Appalachia; Mencher argued for the Taconic Mountains as the source because pebbles of metamorphic rock in the Catskill sequence matched rocks in the Taconics and because proof of granitic debris was lacking in the Catskill rocks. He presupposed at that time that Appalachia contained granitic rocks. Barrell did no thin-section petrography, but concluded from megascopic observation that Appalachia was largely quartzite, gneiss, and granite.

Krynine's meticulous thin-section studies (1940, 1946) led him to conclude, particularly concerning the Oswego and Juniata formations, that (1946, p. 3) "There is no material whatsoever from a hypothetical crystalline, igneous Appalachian" [sic]. He inferred a source from uplifted weakly metamorphosed sediments from the Magog belt as defined earlier by Kay. It was the petrographic data by Stow (1938), Mencher (1939), and Krynine (1940) that provided a major piece of evidence to support Kay's (1951, p. 56) concept for tectonic lands being composed of metasedimentary rock rather than granitic and gneissic basement rock.

A recent summary statement on the problem is that of Pettijohn (1970, p. 2), who uses seven major petrographic reports as a basis for concluding:

> the sandstones of late Ordovician and younger systems . . . contain very little feldspar, are devoid of volcanic debris, are quartz-rich, and contain particles of metaargillaceous rocks such as slate, phyllites, and metaquartzite. From these observations we infer a source land with few, if any, deep-seated plutonic granites and gneisses, little or no volcanic rocks, and composed largely of sedimentary and metasedimentary rocks. Such conclusions seem to rule out either the Canadian Shield with its extensive granites and related rocks or peripheral island arcs of volcanic construction.

The Last Decade

New ideas have been generated in the last ten years or so by the addition of new data on continental-margin structure, by proof of sea-floor spreading, and additional stratigraphic and structural data on sedimentary sequences along the Atlantic margin as a means of evaluating continental drift.

Drake and others (1959) drew attention to the close analogy between Kay's miogeosynclinal-eugeosynclinal couple and the paired inner-trough (shelf) and continental-rise sediments off the present eastern margin of the North American continent. In a sequence of essays and discussions, Dietz and colleagues (Dietz, 1963, 1965, 1968; Dietz & Holden, 1966; Dietz & Sproll, 1968) expanded on this idea, which, as Bird and Dewey (1970, p. 1037) note, was implied but not specially stated earlier by Kay (1945). According to Dietz (1963, p. 314); "... the continental terrace sedimentary wedges are modern miogeosynclines ... and the subjacent continental-rise sedimentary prisms are modern eugeosynclines. ... The ... mountain building cycle is initiated by thrusting probably related to decoupling of a spreading sea floor. ... The continental rise is then thrust, folded, and metamorphosed, forming a eugeosyncline. ... *Island arcs, borderlands, trenches, and tectogenes play no part in this actualistic version*" (italics mine).

Many of Dietz's and his colleague's statements were challenged on sedimentologic, stratigraphic, and geochronologic issues (Hsu 1965 a; Clifford & Henderson 1968; Young 1968). Two fundamental issues in the actualistic version are whether the continental-rise prism is at all similar in composition and provenance to the eugeosynclinal sequence (as seen by Kay), and whether the metamorphic Piedmont rocks of the central Appalachians are the eugeosynclinal segment of the miogeosynclinal and unmetamorphosed folded Appalachian Basin rocks—or part of an earlier phase of geosynclinal deposition and deformation. The latter question was reviewed recently by Pettijohn (1970).

One novel idea introduced by Dietz (1963, p. 314) was that of orogenesis of the Atlantic margin by underthrusting of a spreading sea-floor (sima) at the margin of the continental sialic "raft." Dietz at this time apparently had in mind an Atlantic Ocean with its present size and configuration and was confident in the inexistence of Appalachia as a sizable landmass.

An alternative to the idea that Appalachia was not a sizable landmass is a mechanism whereby a once-existing landmass disappears. Schuchert (1923) thought that Appalachia "foundered," but did not offer a mechanism. Hsu (1965 b) treated this concept and discarded the idea of (1) oceanization of sialic crust and (2) "subcrustal erosion" by convection currents within the mantle in favor of the hypothesis that Appalachia foundered due to phase changes within the mantle beneath it. He proposed that an initial decrease in mantle density permitted an isostatic rise of the crust to form the landmass;

and that supracrustal erosion, plus an increase in density of the mantle during a latter phase, resulted in isostatic subsidence and resultant foundering beneath the Atlantic.

Although it is likely that stratigraphers of the United States have long considered the possibility that Appalachia was a part of Laurasia or Pangaea, I find no written mention of the idea prior to Willard (1939, p. 362). Modern adherents (Bullard et al. 1965; Dietz & Holden 1970; and others) argue for either separation during the Mesozoic or a complicated history of continental separations and collisions (Wilson 1966). Bird and Dewey (1970, p. 1031) offer a specific history of drifting as related to orogenesis in the northern Appalachians. They propose "Late Precambrian to Ordovician through Devonian contraction of a Proto-Atlantic Ocean. This oceanic opening and closing was achieved by initial extensional necking of a single North American African continent and by lithosphere plate accretion, followed by contractional plate loss along a trench or complex of trenches, marginal to the drifted North American continent. . . . Pre-orogenic Appalachian sedimentation models were essentially the same as those found along modern continental margins; that is, shelf/slope/rise/abyss."

By their hypothesis, and also Kay's and Dietz's, almost all non-volcanic detritus in the eastern Appalachian Basin and Piedmont rocks was ultimately derived from the North American craton (that is, Appalachia was composed of detritus derived from the North American craton).

In summarizing the results of the Gander Conference on "North Atlantic—Geology and Continental Drift," Kay (1969, p. 967) notes: "Thus, the similarities in the stratigraphic-structural belts of Newfoundland and the British Isles are so great that the probability of their being accidental is trivial. If the two areas were not originally contiguous at the continental shelves, there must have been continuing belts beneath the ocean."

Allen et al. (1967) and Friend (1969) reviewed the distribution of non-marine rocks of the "old Red" clastic sequence along the borders of the North Atlantic. They conclude that thickness distributions and paleocurrent data are compatible with a pre-drift continental configuration of Bullard et al. (1965).

If the Laurasia configuration of Bullard et al. (1965) is correct, Appalachia may have included terrain within present continental Africa. This possibility has been considered by Wilson (1965, p. 149) and Rona (1970) among others. As noted by Kay (1969, p. 968) "The geology of the African coast is insufficiently known to substantiate that rocks were contiguous with those of the American Atlantic Coast before the opening of the ocean."

Assuming for a moment the existence of a Precambrian to Paleozoic Laurasia or Pangaea, we have at least three possible histories of Appalachia:

1) Appalachia during late Precambrian and Cambro–Ordovician time was largely what is now Africa and its root is preserved, or

2) Appalachia during Precambrian and Cambro-Ordovician time was largely what is now Africa and part of the U. S. Atlantic coastal margin—and only part of its root is preserved, or

3) Appalachia was constructed during middle Paleozoic time from sediments deposited in a narrow trough between what is now Africa and North America and its roots were destroyed during subsequent drifting and sea-floor spreading. A subsidiary problem to the latter hypothesis is the source of detritus comprising Appalachia. The actualistic (analog) hypothesis implies derivation from the North American craton, whereas the stratigraphic-sedimentologic evidence suggests an eastern (African?) source. As noted by Pettijohn (1970, p. 3), facies relations and paleocurrent data suggest an easterly source for the Snowbird Group, Great Smokey Group, Glenarm Series, and Lynchburg Formation in the central and southern Appalachian Piedmont. Until the history of these Precambrian and Early Paleozoic sedimentary rocks and metasedimentary rocks of the Piedmont (King 1970*a*, 1970*b*) can be worked out in more detail, we have too many questions unanswered. Techniques of sedimentary petrology offer considerable hope of providing a solution to the problem of Appalachia.

REFERENCES

Adams, R. W. 1970. Loyalhanna Limestone—Cross-bedding and provenance, *in* Fisher, G. W., et al. (eds.), *Studies in Appalachian geology: central and southern.* New York: Wiley–Interscience, 83–100.

Allen, J. R. L., Dinely, D. L., & Friend, P. F. 1967. Old Red Sandstone basins of North America and Northwest Europe, *in* Oswald, D. H. (ed.), *International Symposium on the Devonian System. Alberta Soc. Pet. Geol. 1,* 69–98.

Allen, J. R. L., & Friend, P. F. 1968. Deposition of the Catskill Facies, Appalachian Region: with notes on some other Old Red Sandstone Basins, *in* Klein, G. deV. (ed.), *Late Paleozoic and Mesozoic sedimentation, northwestern North America. Geol. Soc. America,* Special Paper 106, 1–20.

Barrell, Joseph. 1907. Origin and significance of the Mauch Chunk shale. *Geol. Soc. America Bull. 18,* 449–76.

_____. 1908. Relations between climate and terrestrial deposits. *Jour. Geology 16,* 159–90, 255–95, 363–84.

_____. 1912. Criteria for the recognition of ancient delta deposits. *Geol. Soc. America Bull. 23,* 377–446.

_____. 1913. The Upper Devonian delta of the Appalachian geosyncline. *Am. Jour. Sci.* [4], *36,* 429–72.

———. 1914. The Upper Devonian delta of the Appalachian Geosyncline. *Am. Jour. Sci.* [4], *37*, 87–109, 225–53.

Behre, C. H., Jr. 1927. Slate in Northampton County, Pennsylvania. *Pa. Topog. Geol. Survey, 4th ser., Bull. M9*, 308.

Bird, J. M., & Dewey, J. F. 1970. Lithosphere plate–continental margin tectonics and the evolution of the Appalachian Orogen. *Geol. Soc. America Bull. 81*, 1031–60.

Brown, W. R. 1970. Investigation of the sedimentary record in the Piedmont and Blue Ridge of Virginia, *in* Fisher, G. W., et al. (eds.), *Studies in Appalachian geology: central and southern.* New York: Wiley–Interscience, pp. 335–49.

Bullard, Edward, Everett, J. E., & Smith, A. G. 1965. The fit of the continents around the Atlantic, in *A symposium on continental drift.* Blackett, P. M. S., Bullard, E. G., and Runcorn, S. K. (eds.), *Phil. Trans. Roy. Soc. London 1088*, 41–51.

Cavaroc, V. V., Jr., & Ferm, J. C. 1968. Siliceous spiculites as shoreline indicators in deltaic sequences. *Geol. Soc. America Bull. 79*, 263–72.

Clarke, J. M., & Ruedemann, R. 1962. The Eurypterida of New York. *New York State Museum Mem. 14*, 439.

Clifford, P. M., & Henderson, J. R. 1968. Miogeoclines (Miogeosynclines) in space and time: a discussion. *Jour. Geology 76*, 111–12.

Colton, G. W. 1970. The Appalachian basin—its depositional sequences and their geologic relationships, *in* Fisher, G. W., et al. (eds.), *Studies in Appalachian geology: central and southern.* New York: Wiley–Interscience, pp. 5–48.

Dana, J. D. 1846. On the volcanoes of the moon. *Am. Jour. Sci. 2*, 335–55.

———. 1856*a*. On American geological history. *Am. Jour. Sci. 22*, 303–34.

———. 1856*b*. On the plan of development in the geological history of North America. *Am. Jour. Sci. 22*, 335–49.

———. 1890. Areas of continental progress in North America. *Geol. Soc. America Bull. 1*, 36–48.

Dietz, R. S. 1963. An actualistic concept of geosynclines and mountain building. *Jour. Geology 71*, 314–43.

———. 1965. Collapsing continental rises, an actualistic concept of geosynclines and mountain building; a reply. *Jour. Geology 73*, 901–6.

———. 1968. Miogeoclines (Miogeosynclines) in space and time, a reply. *Jour. Geology 76*, 119–21.

Dietz, R. S., & Holden, J. C. 1966. Miogeoclines in space and time. *Jour. Geology 74*, 566–83.

———. 1970. Reconstruction of Pangaea: breakup and dispersion of continents, Permian to Present. *Jour. Geophys. Research 75*, 4939–56.

Dietz, R. S., & Sproll, W. P. 1968. Miogeoclines (Miogeosynclines) in space and time: a reply. *Jour. Geology 76*, 113–16.

Donaldson, A. C. 1969. Ancient deltaic sedimentation (Pennsylvania) and its control on the distribution, thickness and quality of coals, *in* Donaldson, A. C. (ed.), Some Appalachian Coals and Carbonates. Models of Ancient Shallow Water Deposition. *Field Guide, West Virginia Geol. Survey*, pp. 93–121.

Drake, C. L., Ewing, M. & Sutton, G. H. 1959. Continental margins and geosynclines: the east coast of North America North of Cape Hatteras, in *Physics and chemistry of the earth*. London: Pergamon Press *3*, 110–98.

Ferm, J. C. 1970. Allegheny deltaic deposits, *in* Morgan, J. P. (ed.), *Deltaic sedimentation, modern and ancient*. Soc. Econ. Paleontologists and Mineralogists Spec. Pub. *15*, 246–55.

Ferm, J. C., & Williams, E. G. 1965. Characteristics of Carboniferous marine invasions in Western Pennsylvania. *Jour. Sedimentary Petrology 35*, 319–30.

Friend, P. F. 1969. Tectonic features of Old Red sedimentation in North Atlantic borders, *in* Kay, Marshall (ed.), *North Atlantic geology and continental drift*. Am. Assoc. Petroleum Geologists Memoir *12*, 703–10.

Grabau, A. W. 1913. Early Paleozoic delta deposits of North America. *Geol. Soc. America Bull. 24*, 399–528.

Hsü, K. J. 1965*a*. Collapsing continental rises: an actualistic concept of geosynclines and mountain building: a discussion. *Jour. Geology 73*, 897–900.

———. 1965*b*. Isostasy, crustal thinning, mantle changes, and the disappearance of ancient land masses. *Am. Jour. Sci. 263*, 97–109.

Hunter, R. E. 1970. Facies of iron sedimentation in the Clinton Group, *in* Fisher, G. W., et al. (eds.) *Studies of Appalachian geology: central and southern*. New York: Wiley–Interscience, pp. 101–21.

Hyde, J. E. 1911. The ripples of the Bedford and Berea formations of central Ohio with notes on the paleogeography of that epoch. *Jour. Geology 19*, 257–69.

Kay, G. M. 1951. *North American geosynclines. Geol. Soc. America Memoir 48*, 143.

Kay, Marshall. 1944. Geosynclines in continental development. *Science 99*, 461–62.

———. 1969. Continental drift in North Atlantic Ocean, *in* Kay, Marshall (ed.), *North Atlantic geology and continental drift. Am. Assoc. Petroleum Geologists Memoir 12*, 965–73.

King, P. B. 1970*a*. The megatectonics of continents and oceans, in Johnson, Helgi and Smith, B. L. (eds.), *Tectonics and Geophysics or Eastern North America*. New Brunswick, N.J.: Rutgers University Press, pp. 74–112.

———. 1970*b*. The Precambrian of the United States of America: Southeastern United States, *in* Rankama, K. (ed.), *The geologic systems: the Precambrian*. New York: Interscience, John Wiley and Sons, *4*, 1–75.

Klein, G. de V. 1967. Paleocurrent analysis in relation to modern marine sediment dispersal patterns. *Am. Assoc. Petroleum Geologists Bull. 51*, 366–82.

Krynine, P. D. 1940. Petrology and genesis of the Third Bradford Sand, Pennsylvania. *Penna. State Coll. Min. Inds. Expt. Sta. Bull. 29*, 134.

———. 1946. From the Cambrian to the Silurian near State College and Tyrone. 12th Annual Field Conf. of Penna. Geol., pp. 2–17.

LaPorte, L. F. 1967. Carbonate deposition near mean sea-level and resultant facies mosaic: Manlius Formation (lower Devonian) of New York State. *Am. Assoc. Petroleum Geologists 51*, 73–101.

Lesley, J. P. 1892. Summary description of the geology of Pennsylvania. *Pennsylvania Geol. Survey 2d Rept. 1*, 1–720.

Lesley, J. P., d'Invilliers, E. V., & Smith, A. D. 1895. Summary description of the geology of Pennsylvania. *Pennsylvania Geol. Survey Rept. 3*, Pt. 1, 1629–2152.

McBride, E. F. 1962. Flysch and associated beds of the Martinsburg Formation (Ordovician), central Appalachians. *Jour. Sedimentary Petrology 32*, 39–91.

McIver, N. L. 1970. Appalachian turbidites, *in* Fisher, G. W., et al. (eds.), *Studies of Appalachian geology: central and southern.* New York: Wiley–Interscience, pp. 69-82.

Meckel, L. D. 1970. Paleozoic alluvial deposition in the central Appalachians: a summary, *in* Fisher, G. W., et al. (eds.), *Studies of Appalachian geology: central and southern.* New York: Wiley–Interscience, pp. 49–68.

Mencher, Ely. 1939. Catskill facies of New York State. *Geol. Soc. America Bull. 50*, 1761 –94.

Pelletier, B. R. 1958. Pocono paleocurrents in Pennsylvania and Maryland. *Geol. Soc. America Bull. 69*, 1033–64.

Pepper, J. F., DeWitt, W. Jr., & Demarest, D. F. 1954. Geology of the Bedford Shale and Berea Sandstone in the Appalachian Basin. *U.S. Geol. Survey Prof. Paper 259*, 111 pp.

Pettijohn, F. J. 1957. *Sedimentary rocks*, 2nd ed., New York: Harper and Brothers, 718 pp.

———. 1962. Paleocurrents and paleogeography. *Am. Assoc. Petrol. Geol. Bull. 46*, 1468–93.

———. 1970. Introduction, *in* Fisher, G. W., et al. (eds.), *Studies in Appalachian geology: central and southern.* New York: Interscience, pp. 1–4.

Potter, P. E., & Pettijohn, F. J. 1963. *Paleocurrents and basin analysis.* Berlin: Springer–Verlag, 296 pp.

Rickard, L. V. 1962. Late Cayugan (Upper Silurian) and Helderbergian (Lower Devonian) stratigraphy in New York. *New York State Mus. and Science Service, Bull. 386*, 157 pp.

Rona, P. A. 1970. Comparison of continental margins of eastern North America at Cape Hatteras and northwestern Africa at Cape Blanc: discussion. *Am. Assoc. Petroleum Geologists Bull. 54*, 2216–18.

Ruedemann, R. 1897. Evidence of current action in the Ordovician of New York. *Am. Geologist 19*, 367–91.

Schuchert, Charles. 1910. Paleogeography of North America. *Geol. Soc. America Bull. 20*, 427–606.

———. 1916. Silurian formations of southeastern New York, New Jersey, and Pennsylvania. *Geol. Soc. America Bull. 27*, 531–54.

———. 1923. Sites and names of North American Geosynclines. *Geol. Soc. America Bull. 34*, 151–229.

Schuchert, Charles, & Dunbar, C. O. 1941. *Textbook of geology:* part 2, historical geology, 4th ed. New York: John Wiley, 551 pp.

Schwab, F. L. 1970. Origin of the Antietam Formation (Late Precambrian ?–Lower Cambrian), Central Virginia. *Jour. Sedimentary Petrology 40*, 354–66.

Smith, N. D. 1970. The braided stream depositional environment; comparison of the Platte River with some Silurian clastic rock, north central Appalachians. *Geol. Soc. America Bull. 81*, 2993–3014.

Stowe, M. H. 1938. Conditions of sedimentation and sources of the Oriskany Sandstone as indicated by petrology. *Am. Assoc. Petroleum Geologists Bull. 22*, 541–64.

Sutton, R. G., Bowen, Z. P. & McAlester, A. L. 1970. Marine shelf environments of the Upper Devonian Sonyea Group of New York. *Geol. Soc. America Bull. 81*, 2975–92.

Swartz, C. K., & Swartz, F. M. 1931. Early Silurian formations of southeastern Pennsylvania. *Geol. Soc. America Bull. 42*, 621–62.

Van Houten, F. B. 1954. Sedimentary features of Martinsburg slate, Northwestern New Jersey. *Geol. Soc. America Bull. 65*, 813–18.

Walker, R. G., & Harms, J. C. 1971. The Catskill delta; a prograding muddy shoreline in central Pennsylvania. *Jour. Geology 79*, 381–400.

Walker, R. G. 1971. Non-deltaic depositional environments in the Catskill clastic wedge (Upper Devonian) of central Pennsylvania, *Geol. Soc. America Bull. 82*, 1305–26.

Wanless, H. R., et al. 1970. Late Paleozoic deltas in the central and eastern United States, *in* Morgan, J. P. (ed.), *Deltaic sedimentation; modern and ancient.* Soc. Econ. Paleontologists and *Mineralogists Spec. Paper 15*, 215–46.

Willard, Bradford. 1939. The Devonian of Pennsylvania. *Pennsylvania Geol. Survey Bull.* G19, 481 pp.

Williams, E. G., & Ferm, J. C. 1964. Sedimentary facies in the lower Allegheny rocks of western Pennsylvania. *Jour. Sedimentary Petrology 34*, 610–14.

Williams, H. S. 1897. On the southern Devonian formations. *Am. Jour. Sci.* [4], *3*, 393–403.

Willis, Bailey. 1902. Paleozoic Appalachia or the history of Maryland during Paleozoic time. *Maryland Geol. Survey* Part IV, *1*, 23–93.

Wilson, J. T. 1965. Evidence from ocean islands suggesting movements in the earth, *in* Blackett, P.M.S., Bullard, E. C., and Runcorn, S.K., (eds.), *A symposium on continental drift. Phil. Trans. Roy. Soc. London 1088*, 145–67.

――――. 1966. Did the Atlantic Ocean close and reopen. *Nature 211*, 676.

Wolcott, C. D. 1891. Correlation papers, Cambrian. *U.S. Geol. Survey Bull. 81*, 363–69.

Wolff, M. P. 1965. Sedimentologic design of deltaic sequences, Devonian Catskill complex of New York· (abs). *Am. Assoc. Petroleum Geologist Bull. 49*, 364.

Yeakel, L. S., Jr. 1962. Tuscarora, Juniata, and Bald Eagle paleocurrents and paleogeography in the central Appalachians. *Geol. Soc. America Bull. 73*, 1515–40.

Young, G. M. 1968. Miogeoclines (Miogeosynclines) in space and time; a discussion. *Jour. Geology 76*, 116–18.

Chapter Five

Carbonate Petrography in the Post-Sorbian Age

ROBERT L. FOLK

INTRODUCTION: H. C. SORBY AND CARBONATE PETROGRAPHY

Carbonate petrography has come a long way since the turn of the century; nevertheless, in re-reading the works of Henry Clifton Sorby, one is amazed that it took so long to build upon the excellent foundation that he had laid—a foundation virtually ignored for seventy years because petrographers sniffed at sedimentary rocks as being mineralogically uninspiring. Paleontologists and stratigraphers in pursuit of new species and formation names ignored the depositional environment of the rocks, and oil companies had not yet found much oil in limestones or much sense in studying sediments instead of anticlines.

H. C. Sorby was born in Sheffield, England, in 1826 (Folk 1965a) and, supported by inherited wealth, was able to devote his entire life to independent research, following wherever his curiosity led. He never married, never held a geologic job, never attended a University, and was a supreme individualist. Besides his multitude of other accomplishments and discoveries, he was the first geologist to make a systematic study of paleocurrents in clastic sediments, and he also founded the science of metallography, studying components, recrystallization, and deformational textures in metals.

With all our recent advances, I have not yet been able to find one erroneous statement in Sorby's work on carbonate rocks. Everything he said is still valid today, even if then but dimly perceived. Reading Sorby is almost like reading the Bible; if one searches hard enough, he can find a hint on almost any point desired. His 1851 paper on the petrography of chertified limestone was the first paper published in *any* field of petrography, and pre-dated the birth of igneous petrography by twelve years (Folk 1965a, pp. 44, 93). Sorby (1853) did the first quantitative petrography, getting the percentage composition of a carbonate rock (by tracing images on paper, then cutting and weighing the pieces), and (1855) strongly affirming that thin sections were the only real means of finding out the true composition and structure of limestones. In his landmark paper of 1879, well worth reading today by every petrographer, he pointed out that most limestones consist of mechanically transported grains, including fossil fragments, oolites, reworked pieces of older carbonate sediment, and structureless pellets. These are the same four major framework components recognized today.

Although the microscopic structure of modern shells had been worked up in great detail by Carpenter (1844, 1847), Sorby was the first to determine which shells were aragonite and which were calcite, and to realize the effect that structure and mineralogic composition had upon abrasion, disintegration, and preservation of shells in the fossil record. Sorby (1862) determined that calcite shells were stable, but aragonite shells were selectively dissolved out upon weathering; and that aragonite could invert to calcite by heating, prolonged soaking in water, or decay of the intimate organic binding of the crystallites. He showed that aragonite shells could change to calcite by either (1) change of molecular state while still solid, to a mosaic of unoriented calcite crystals—in fact he accomplished this in a boiler, thus doing the first experimental work on diagenesis; or (2) by complete dissolution of the aragonite, leaving a cavity which could be filled much later by calcite crystals projecting inward as in a geode or vein-filling. Because of the nature of their crystal units, different shells would break down physically into mud particles of different sizes and shapes and would show very different rates of durability and chemical survival (1853, 1879). Oolites were formed in agitated, probably warm, waters, mainly of aragonite prisms accreted mechanically and lying tangentially on the surface of the oölite. Concentric bands were caused by varying rates of chemical precipitation or abrasion and radial calcite formed by alteration of the aragonite. Fragments of contemporaneous carbonate sediment or even of older lithified limestone could be incorporated into limestones. Sorby (1861) first showed that coccoliths were organic and a major component of chalk, and he compared the origin of chalk to that of deep-sea ooze.

Carbonate grains of several types could then be deposited by waves or currents like modern carbonate shell-sands, forming beds of porous clastic sediment, and the pores could then be filled by precipitation of calcite cement, much of it derived from alteration of aragonite skeletons. The cement took the form of singly-crystalline overgrowths upon and within echinoderm fragments, or of radial fibers on trilobites or ostracods, in both instances as optically-continuous overgrowths. In other limestones, laid down in more tranquil water, the transported grains were embedded in a matrix of carbonate mud. Sorby thought there were at least four origins of carbonate mud: (1) physical abrasion of fossils, (2) organic breakdown (rotting) of fossils, (3) chemical precipitation, and (4) abrasion of older limestones. He favored an origin by disintegration of carbonate skeletons, giving muds of characteristic crystal size and shape, but he thought aragonite muds could change to calcite and obliterate evidence of origin. The dolomites he studied formed by replacement of limestones, as shown by transection of fossils. Sorby's final paper (1904) on the Funafuti boring ended over half a century of brilliant work on carbonate petrography.

In this brief review of advances since 1900, I have covered only subjects that particularly interest me. An entire book could be written on the new techniques that were not available to Sorby.

I have also avoided a systematic survey of the many realms of sedimentation now known to produce carbonates (Friedman 1969), from caliches, playa lakes, and aeolianites, to beaches, tidal flats, and, finally, reefs, open shelves, and turbidites. I have concentrated on the petrography of carbonate rocks but included data on modern carbonate sediments.

MAJOR COMPONENTS OF CARBONATES: CLASSIFICATION

Classification these days is a nasty word and classifiers are regarded as the "fussy old maids" of the profession. But it should not be forgotten that classification is an inevitable development of the growth of knowledge about a subject. Before a successful classification can be developed, one must know the field of variation of the subject concerned and must perceive genetic relationships and important principles. A classification is merely the tip of the knowledge iceberg, summarizing in a few short syllables the vast bulk of the subject that lies below the surface. This is true whether one is classifying trilobites, books in a library, or Picasso paintings.

Grabau (1903) made the first comprehensive classification of carbonate rocks. He divided limestones into (I) chemical (oölitic, tufa, etc.), (II) organic, and (III) clastic types, the latter divided on grain size into calcirudites, calcarenites, and calcilutites. Reading standard texts of the 1920's through the 1940's, one finds an absolute regression in carbonate knowledge from Sorby's time, save for the efforts of a few outstanding individuals whose labors never were integrated into textbooks (e.g., Pia 1933; Cayeux 1935). Discussion of limestones in most texts dealt with chemistry (i.e., where did one draw the line between limestone and magnesian limestone? or how much clay content made it into a marl?); or organic composition, with a dreary catalogue of dozens of paleontological names, such as crinoidal limestone, brachiopod limestone, nummulite limestone, etc. Other classifications were based on a "random-walk" principle, naming every type that happened to come to hand, such as oölitic or pisolitic limestone, algal limestone, coquina, tufa, spergenite, lithographic limestone, travertine, glauconitic or pyritic limestone, etc.

A few geologists used a grain-size scale for limestones, such as those of Hirschwald (Howell 1922) or DeFord (1946). Most workers, including Grabau, paid little or no attention to the nature of the matrix and to its vital role in interpreting the origin of carbonate rocks. Sparry calcite was generally thought to be the result of recrystallization of lime mud, basically a super-diagenetic or even metamorphic effect, hence of no value in deciphering the depositional environment. However, Cayeux (1935) thought some clear calcite was probably cement, not necessarily the product of metamorphism or recrystallization. Hatch, Rastall, and Black (1938) realized that calcite mud-stone must have accumulated in a sheltered environment and that clear calcite could be a pore-filling cement, a fact already known by Sorby in 1879. The slowly dawning recognition of the true meaning of the intergrain material (calcite spar vs. calcite mud) was the key that unlocked the door to interpretation of limestones in terms of depositional environment, hydrodynamic energy, and facies distribution.

Pettijohn (1949, 1957) recognized the role of depositional environment and established two main types of limestones. Allochthonous rocks were made up of transported particles, such as fossils and oölites, tended to be sorted, and have a crystalline calcite cement. Autochthonous limestones were not transported and included fossils in a lime-mud matrix, pelagic limestones, reef rocks, etc.

Folk developed the principles of his classification in 1948 during the course of his thesis under P. D. Krynine on the Beekmantown rocks of central Pennsylvania (Folk 1952). Inspired by Krynine's treatment of sandstones in terms of the three textural end-members of grains, clay matrix, and chemical cement, Folk applied the identical concepts to carbonates and realized that they also consisted of mechanically transported grains, carbonate mud matrix, and pore-filling sparry calcite cement. The Beekmantown rocks were relatively simple, showed very little recrystallization, and were easy to classify on this sandstone-analogy, hydraulic-energy basis. Framework grains (or allochems) were divided into four types: fossils, oölites, intraclasts (reworked fragments of carbonate sediment), and tiny pellets (assumed to be fecal). After teaching the system for four years, Folk presented the classification at the American Association of Petroleum Geologist's meeting in St. Louis in 1957. At about the same time, the same society saw the need for a carbonate research program and set up the Carbonate Research Subcommittee; large oil fields were being found in carbonate reservoirs. Bramkamp and Powers (1958) and Folk (1959*a*, 1959*b*) then brought out their complete classifications; both were based on similar basic concepts but each had its own terminology. In March 1959 the subcommittee decided to bring all the worms out of the woodwork by proposing that an entire SEPM session be devoted to carbonate classification systems. Illustrating the importance of oil company research in carbonates, all but one of the papers in the subsequent volume (Ham 1962) originated from the petroleum industry. The carbonate research boom was suddenly on! Hordes of oil company geologists were now descending on tropic isles; research labs and carbonate field trips sprang up like toadstools. The Shell Development Company Research Laboratory led the pack in both intensity and quality of research.

Of the classifications proposed in this volume, that of Dunham (1962) became most widely used. He emphasized depositional texture without proposing any formal terminology for allochem types. Folk's micrite and Dunham's lime-mudstone both with less than 10 percent allochems are synonyms. Dunham divided Folk's biomicrite, oömicrite, etc. into two useful divisions depending upon whether the allochems were suspended (meaning they had to have been deposited in a mud matrix, regardless of whether it had later totally recrystallized to sparry calcite or not), or were grain-supported. These two types were called wackestone and packstone. Finally, Folk's biosparite, oosparite, etc. were synonymous with Dunham's grainstone, except that Dunham drew the limit at less than 1 percent mud.

By now dozens of carbonate classification systems exist, nearly all of them based on similar hydraulic philosophies but using slightly different boundary lines and terms.

Allochems

The main transported constituents of carbonate rocks had been known since the mid-nineteenth century, and many of them had been exquisitely pictured in the book of Cayeux (1935). Folk (1952, 1957, 1959*a*) organized them formally into a quantitative classification, and coined for them the inclusive term allochem: out-of-the-ordinary chemical precipitates which were aggregated and transported. Four types of allochems were recognized in marine limestones: (1) fossils (skeletal grains), (2) oölites (including pisolites), (3) fragments of reworked carbonate sediment, termed "intraclasts," and (4) tiny, uniformly sized ovoid micritic pellets, thought to be of fecal origin. Later classifications (Ham 1962) considerably expanded this grouping, but the framework remains valid for nearly all limestones.

Intraclasts

The term "intraclast" was coined by Folk (1952, 1957, 1962*c*) to embrace all types of reworked, penecontemporaneous carbonate sediment, representing fragments from fine-sand size (0.2mm) on up to boulders, and originating from within the depositional environment. Thus an intraclast is a *clast* derived from *within* the formation; the word has its roots in the term "intraformational conglomerate" of Walcott (1894) who recognized reworked, semi-consolidated carbonate and pebbles in Paleozoic carbonates.

Following Walcott, Field (1916) classified various kinds of intraformational conglomerates and observed them forming in the intertidal area of the Bahamas, aided by mudcracking and wave attack on small cliffs (Field 1919, 1931). In Paleozoic rocks, carbonate mud clasts are typically discoidal (derived from rip-up of laminated pellet-limestones), and they may lie in any orientation from horizontal to dipping to near-vertical (McKee & Resser 1945). Nason (1901) first used the term "edgewise conglomerate" for these steeply dipping clasts, and Stose (1910) thought they formed by wave erosion in shallow water, then were shuffled about on beaches like shingles; Brown (1913) proposed an origin by slumping. Field (1916) thought edgewise conglomerates were intertidal, and Folk and Robles (1964) proposed that the vertically oriented clasts formed at the high-wave swash-mark during storms, by observing similarly oriented coral sticks on storm-built wave ridges.

Sand-size intraclasts in modern environments were first described in detail by Illing (1954) in the Bahamas, who called them "grapestones," because they were made of weakly cemented aggregates of pellets with botryoidal outer surfaces. Since then, grapestones have been found to be a very common component of modern carbonate sediments in most areas where there is an abundance of carbonate mud, and where the mud can be reworked by waves or currents. Grapestone originates as a sediment of loose pellets which become bonded, usually into a thin, very fragile surface crust in shallow depths. The bonding agent is probably organically inspired (Purdy 1963). Break-up of the crust by storm waves or burrowers produces a "grapestone" sand, which can then be redeposited into beaches, underwater bars, etc. Beales (1958) recognized the common occurrence of grapestones in Paleozoic limestones.

Intraclasts can form in a great many ways, and at various stages of consolidation (Folk 1962c). Several super-detailed classifications of intraclasts have been proposed (Wolf & Conolly 1965; Wilson 1967). A problem is the distinction between true intraclasts of contemporaneous sediment and fragments of much older, lithified carbonate rock derived from erosion of ancient limestone outcrops (making the terrigenous rock known as "calclithite," Folk 1959a). Wolf (1965a) and Chanda (1967) have proposed "limeclast" for all carbonate clasts, dividing them into "extraclasts" (derived from outside the basin), and "intraclasts" (from within the basin). A further problem is that fossils or oölites transformed into micrite by algae (?) resemble intraclasts. According to Beales's (1961) principle of homeomorphy, given enough abrasion and degradation, intraclasts of many origins may eventually come to look alike.

Pellets

Pellets are the most controversial of the nonskeletal components. Those petrographers whose experience is in Lower Paleozoic limestones find them abundant, easy to recognize, and very important as rockformers; those who have worked in Mesozoic or younger carbonates seldom encounter them and find difficulty in identifying them.

As originally used by Hatch, Rastall, and Black (1938), pellets were distinct in their small size (about 0.1 mm), rounded to ovoid or rodlike shape, homogeneous micritic composition, and superbly good sorting. Folk (1959a, 1962c) drew an arbitrary size cutoff at 0.2 mm between pellets and intraclasts and considered that even though the term was basically descriptive, nevertheless, most of the pellets were fecal deposits of invertebrates, like those described by Moore (1939).

Modern carbonate muds contain an abundance of fecal pellets. Most conspicuous are rod-like creations in the 0.5-1 mm range, but fecal pellets of such a size are very rare as significant rock formers in ancient rocks, though scattered nests of these large pellets occur. Cloud's (1962) description of the superabundant 0.03-.1 mm fecal pellets in the Andros Island muds provided an exact analogy of the pellets so abundant in Lower Paleozoic rocks (e.g., Beales, 1958, 1965; Folk, 1962a). Pellets, however, can also form within algal mats, presumably by nonfecal processes of crystallization.

"Grumeleuse" structure of some authors probably represents merged or recrystallized pellets whose outlines have faded. Many muds are obscurely pelleted, and it is obvious that pellets may be excreted or diagenetically altered to varying degrees of fossilizable coherence, much as the difference between the muck in a cow stable and a pile of deer droppings.

Skeletal Grains or Fossils

The early work of Bøggild (1930) was the first advance on the foundation laid out by Carpenter (1844, 1847) and Sorby. Johnson (1951) rapidly became the "Bible" of those petrographers trying to identify bits of fossils in thin section. The explosive knowledge about fossil algae was fathered by the many articles of Johnson, and summarized in his book (1961). Majewski (1969) has made further strides in the identification of skeletal fragments.

Recent emphasis has been on study of the more intimate chemistry and mineralogy of fossils. Chave (1954a) examined the Mg content as the basis for a geologic thermometer, and he also (1964) investigated the differing abrasion rates in the laboratory, as did Folk and Robles (1964), Stoddart (1964), and Force (1969) in natural environments. Swinchatt (1965) reviewed modes of biologic breakdown of skeletal components. In the last few years detailed studies have been made of the mechanics of expulsion of Mg from magnesian-calcite of crinoids or coralline algae (Schroeder 1969), and the exact distribution of trace-elements within the skeletons as revealed by electron probe. Scanning electron microscopy has yielded much data on micro-architecture. Diagenetic alteration in Holocene skeletons has become a fascinating field (Purdy 1968).

Oolites and Pisolites

These curiously uniform, nearly spherical, concentrically coated particles have excited interest for centuries (see excellent review by Brown 1914), and many theories have been proposed to explain their origin. Vaughan (1914) and Bucher (1918) even favored an origin by in situ crystallization of gelatinous $CaCO_3$. However, intensive work in the Bahamas from the 1950's onward reestablished the essential correctness of Sorby's opinion that they formed through a high rate of precipitation in shallow, agitated, warm waters; most occur in a milieu of strongly shifting currents such as tidal channels, shelf-edge sand bars, etc. (Illing 1954; Carozzi 1960; Kendall 1969). Newell, Purdy, and Imbrie (1960) using C^{14} dating demonstrated the Holocene age of the Bahamian oolites and quantified their relation to shallow water.

The role of algae—whether destructive through boring, neutral and merely entrapped, or constructive through precipitation of coatings—remains just as controversial today as it was in the nineteenth century (Wethered 1890; Kalkowsky 1908; Brown 1914; Purdy 1968). It is obvious that algae can play all three roles with varying degrees of success. They make a particularly important contribution in the low-energy, asymmetrical oölites (Freeman 1962; Davis 1966). As for the inorganic needles that make up the tangential coats forming the bulk of the oölites, it has not yet been conclusively shown whether the needles crystallize in that orientation on the surface or are picked up as floating "strays" and adhere mechanically.

Pisolites represent much more of a problem today than they did fifty years ago. Many marine and lacustrine limestones were known to contain pisolites, and these were thought to be algal by analogy with the layered or columnar stromatolites which had similar occurrence and microstructure. Yet pisolites also were known to grow, probably inorganically, in the soil—e.g., the ferruginous or aluminous pisolites of lateritic soils and bauxites, or the calcareous pisolites of caliche soils. Dunham (1965, 1969) and Thomas (1965, 1968) simultaneously proposed the shocking idea that the famous pisolites in the Permian Capitan Reef of west Texas–New Mexico were not algal pisolites, but instead represented a fossil caliche soil zone. This fascinating hypothesis put the fever of doubt upon all pisolite-bearing limestones in the stratigraphic column and led to intensive restudy (Kendall 1969). However, such pisolite zones are a characteristic facies of many reef tracts, and there are so many that it is hard to imagine them as all being exposed to subaerial weathering and vadose circulation. From now on, all pisolitic zones must be looked at with both hypotheses in mind. Pisolites may be (a) transported or (b) in situ; (1) algal-generated or (2) inorganic; (I) formed in shallow marine environment, (II) a lacustrine or fluvial environment, or (III) a soil environment. I think that all possible combinations of these three factors can occur, and we lack good simple criteria to tell these possibilities apart.

Textures of Allochems

Carbonate sediments, like terrigenous ones, run the textural gamut from nearly pure muds through muddy angular sands or clean well-rounded sands, to coarse gravels. Mainly, the texture results from the physical energy of the depositional environment, though in carbonates the relationships are more complex, as pointed out by Swinchatt (1965), and we cannot expect a one-to-one correspondence between the textures of terrigenous sediments and those of carbonates. There is confusing textural contribution from nearly in situ organisms, such as coral sticks breaking off and falling into a mud bottom. Organic fragments may be so rapidly supplied as to overwhelm the sorting or rounding capability of waves or currents. Despite these difficulties quantitative statistical evaluation of the mean grain size, mud content, rounding and surface features of carbonates yields useful information on the depositional environment which can be combined with evidence from gross geometry, sedimentary structures, and fossils.

A considerable body of data was collected by earlier workers on the grain size of carbonate beach sands, using crude grain-size measures. Few conclusions were made, other than that beach sands were "well sorted." Ginsburg (1956) used the content of material finer than 3ϕ (0.12 mm) to help identify south Florida environments. Carozzi and colleagues made a great many quantitative thin-section studies utilizing grain size (clasticity index) of ancient limestones, and plotting detailed stratigraphic logs under the assumption that coarser maximum clast size indicated shallower water, (e.g., Carozzi & Lundwall 1959).

Folk (1962*b*, 1967*a*) and Folk and Robles (1964), studied carbonate textures in the Yucatan region in an attempt to discriminate beach from subtidal sediments and assess wave vigor as a rounding process. They found that carbonate beach sands had the same sorting values as terrigenous beach sediments, and that sorting values of about $\sigma_I = 0.3 - 0.6\phi$ characterized surf-laid carbonate sediments over a 1500-fold size-range from 3ϕ to -8ϕ (fine sand to boulders). Submerged sediments were characterized by poorer sorting and more non-normal size distributions, with a σ_I value of about 0.8ϕ best separating beach from subtidal sediments. For both environments, the size-sorting relation was sinusoidal, depending on sediment modes.

Maxwell, Day, and Fleming (1961), and Maxwell, Jell, and McKellar (1965) studied grain size in the Australian Great Barrier Reef. Hoskin (1963) differentiated nineteen environments on the basis of grain size in Alacran Reef, and, though there was considerable overlap, distinct groupings emerged. Jindřich (1969) discriminated Florida carbonate environments by textural analysis. Maiklem (1970) did a magnificent study of the grain size of shallow sands about the Australian Great Barrier Reef, showing grain-size parameter changes with distance from reefs.

The scale of textural maturity is applicable to carbonate sediments as well as terrigenous ones (Folk 1962c, 1967a). For example, in Isla Mujeres, immature (mud-rich) carbonate sediments occur in a lagoon, submature sands (clean but poorly sorted) lie in a strait swept by unidirectional currents, mature sands (well-sorted but subangular) are found on protected beaches, and supermature sediments (well-sorted, rounded, and polished) dominate exposed, high-energy beaches.

Pilkey, Morton, and Luternauer (1967) and Fosberg and Carroll (1965) measured roundness in carbonate sediments.

Swinchatt (1965) lists the cautions to be used in interpreting carbonate size data. For example, he found the best sorted sediments in Florida reef tract to be further offshore, while nearer shore sediments were muddy and more poorly sorted because they were trapped in a thick baffle of grass. Such reversals must always be guarded against; the winnowing and sorting capability of an environment depends on many things beside depth of water, such as the location of threads of strong current, baffling by plants or thickets of coral or crinoids, binding by organic mats or slimes, rate of supply of carbonate mud, etc.

Textures, like anything else, cannot be used alone. Typical sedimentary structures and sand body geometry of carbonate sediments have been described in many papers, e.g., Imbrie and Buchanan (1965), Klein (1965), and Ball (1967).

ORIGIN OF CARBONATE MUD

The abundance of carbonate mud in the Bahamas was observed by L. Agassiz in 1851; and Dana (1851) had studied mud in lagoons of Pacific atolls, attributing it to abrasion of carbonate skeletal grains on the beaches and to pulverization by browsing of fish. Sorby, as noted previously, had already proposed four of the five mechanisms invoked today, but also favored the skeletal-disintegration origin.

In 1892 Dall suggested an origin by chemical precipitation through heating, agitation, and loss of CO_2, and this became the leading theory for over half a century; among the strong advocates were Vaughan (1914, 1917), who pictured aragonite needles; Johnston and Williamson (1916); Gee (1932), who produced natural-looking aragonite needles in the laboratory; and Black (1933). Bacterial precipitation had a short-lived vogue; it was first proposed by Drew (1914), favored by Bavendamm (1931, 1932) but Lipman (1929) thought that bacteria were not abundant enough. Wood (1941) proposed that much mud was produced by blue-green algae (algal dust). Small amounts of carbonate mud may be derived from erosion of older limestones or caliche crusts and carried to the sea by streams or wind. Chalks seem to be mainly made of disintegrated calcitic coccoliths, thus explaining their lack of induration.

Lowenstam (1955) first suggested that the disintegration of calcified green algae could produce argonite mud. In 1957 Lowenstam and Epstein used the similarity of oxygen isotope ratios in calcified codiacean algae and argonite mud from the area west of Andros Island, Bahamas, to argue that much of the mud came from the algae. Cloud (1962), Wells and Illing (1964), and Milliman (1967) staunchly defended the chemical origin of carbonate mud. Calcite muds of chemical origin are probably rare, because magnesium ions in sea water may poison the calcite crystal surface and inhibit nucleation (Bischoff 1968). Rapid precipitation, higher supersaturation or evaporation also favor aragonite crystallization instead of calcite. Organic coatings may inhibit nucleation and crystallization of either form of $CaCO_3$ (Chave 1965).

Most recent workers now favor the disintegration of algal or invertebrate skeletons as the main source of modern carbonate muds, returning to the century-old ideas of Dana and Sorby (Neumann 1965; Mathews 1966; Stockman et al. 1967; Hoskin 1968). A great difficulty in distinguishing the disintegrational vs. the chemical origin is that aragonite needles produced chemically (Gee 1932) and those liberated by the rotting of green algae are practically indistinguishable, even by the electron microscope (Cloud 1962). Comprehensive reviews of this problem have been made by Revelle and Fairbridge (1957) and Purdy (1963). As with all other controversies, there is no doubt that each of the mechanisms of mud-production proposed operates somewhere, but the problem of assessing their relative contributions in specific areas persists. However, conditions may have been so different in the past that the origin of modern muds may not be too germane to the interpreting of, for example, the widespread Ordovician micrites.

DOLOMITE

Most of the basic petrographic facts on the origin of dolomite were well-established before 1920; the only elements lacking were examples of widespread Holocene dolomite and a sound chemical theory to explain its formation. In the 1920's it was not possible to form dolomite in the laboratory under anything like natural conditions; now, fifty years later, we are still unable to synthesize it easily.

As early as 1843, Dana, from studies of Pacific atolls, postulated that dolomite formed by the reaction of sea water containing Mg^{++} with calcium carbonate sediments, and Skeats (1905, 1918) contended that dolomitization occurred early, in warm seas and in waters shallower than 150 feet. Hunt (1859) had pointed out the association of dolomite with evaporites and postulated that dolomite would form if gypsum were precipitated first to raise the Mg:Ca ratio, an idea that did not reappear for 100 years. Cullis (1904) made a brilliant study of the complex paragenesis of dolomite at depths greater than 638 feet in the Funafuti Atoll boring, and both he and Skeats (1905) showed euhedral crystals of primary dolomite lining cavity walls, correctly concluding that the mineral had formed directly by precipitation not by replacement.

Landmark papers by Steidtmann (1911, 1917) and Van Tuyl (1916) established that the two most common kinds of dolomite could be separated by a crystal-size boundary of $10-20\mu$: (1) the extremely finely crystalline dolomite muds, variously known as "dense," "primary," "aphanitic," "early," or "supratidal," dolomite, that occur characteristically in thin, widespread beds which show ripplemarks, mudcracks, and erosional features and lacking ghost limestone structures or fossils; and (2) "crystalline," "secondary," "sugary," "late," "replacement," or "reflux," dolomite often occurring irregularly, replacing limestone with abundant evidence of metasomatism. Steidtmann recognized the first type as a chemically precipitated marl and the second type as a later replacement along zones of greater permeability. Sander (1936) also stressed the importance of primary, cavity-filling dolomite and laminated dolomite muds, and even described rolled, detrital grains of dolomite. Stout (1944) noted that in Ohio, shale was associated with limestone beds, while dolomite beds were clay-free, and concluded that freshwater influx caused calcite to precipitate as well as bringing in clay, while dolomite formed in clearer waters of higher salinity.

The interpretation of the origin of dolomite was hampered by the absence of recent examples. However, as early as 1911 Wade described rhombs that he called dolomite in Holocene lagoonal marls along the desert Red Sea coast of Egypt, and he believed they were being precipitated in the hot, supersaline waters; he postulated a similar evaporite-pan origin for Triassic dolomite beds in England. Mawson (1929) also found modern dolomitic marl forming in the Coorong Lagoon of South Australia. These occurrences were ignored until Alderman and Skinner (1957) restudied the Coorong and set off an epidemic of discoveries of recent dolomite in tropical, semi-arid, supratidal or very shallow lagoonal areas. The spotlight flitted almost yearly from one "hot area" to another; the Coorong yielded more than its quota of papers (e.g., von der Borch 1965), then came Florida Bay (Shinn 1964), the Bahamas, where "supratidal" crusts and mud flats of dolomite came into vogue (Shinn, Ginsburg & Lloyd 1965), the sebkhas of the Persian (Arabian) Gulf (Illing, Wells & Taylor 1965), the Netherlands Antilles (Deffeyes, Lucia & Weyl 1965), and Pacific atolls (Berner 1965). "Supratidal" dolomite rapidly became the dogma of the decade. Next came two rival hypotheses: one, the "downsoakers" (e.g., Butler 1969) who wanted to flood the mudflats at abnormally high tides and let the evaporating sea water soak down, forming dolomite as it became more concentrated; and the other "upsuckers" (e.g., Hsü & Siegenthaler 1969) who wanted to draw sea water up through the mud by evaporation and capillary suction.

Coarsely crystalline dolomite was long recognized as a replacement, but little attention was paid to the mechanism until King (1947) and Krynine

(1957) proposed and Adams and Rhodes (1960) popularized the reflux idea: heavy brines enriched in magnesium from hypersaline lagoons moving down through porous sediments below and dolomitizing the sediments or reefs in their paths (Skeats had already intimated this idea in 1903). Experiments by Deffeyes, Lucia, and Weyl (1965) apparently supported this mechanism, but recently Murray (1969), and Hsü and Siegenthaler (1970) have seriously questioned the reflex hypothesis.

The dogma of the decade: supratidal and reflux dolomite—has become too one-sided. Though it is true that the great majority of dolomites do indeed form under tropical, semi-arid, high salinity, shallow to supratidal environments (already known in the 1800's), yet there are also other extensive occurrences. For example, dolomite is common in some lacustrine environments (Klähn 1928), and even in low-saline glacial Lake Agassiz (Sherman & Thiel 1939) or in caliche and soils (Sherman 1947; Sherman, Schultz & Alway 1962) and can form by meteoric reshuffling of ions in magnesian-calcite sediments (Land & Epstein 1970). It also is found in deep-sea sediments or in the bottoms of hypersaline bays (Behrens & Land 1970) or in shallow, nearly normal sea water (Atwood & Bubb 1970). There are great thicknesses of dolomite that form by replacement; some unknown proportion may have formed by "reflux" from down-soaking evaporite pans, but in my opinion their volume is rather small. Much coarser dolomite probably forms from connate waters not associated with evaporites and much may be very late, from hydrothermal waters or from solutions traveling along faults (Friedman & Sanders 1967; Spörli, 1968).

In some examples the Ca/Mg of the dolomite may be useful in determining the mode of origin, time of origin, or climatic regime (Fuchtbauer 1965). Marschner (1968) suggested that the variations of Ca/Mg in Triassic dolomites were related to paleosalinities.

Texturally, dolomite rocks show about the same series of lithologic types as limestones. Dolomite muds (dolomicrites), accumulate in low-energy environments such as lagoons, supratidal flats, or bays. Dolomitic algal mounds and dolomitic intraclasts bear the same environmental significance as their limestone analogues. No primary dolomite oölites are yet known to me, though it is very common for dolomite to replace calcite/aragonite oölites. Dolomite formations thus can be subjected to paleogeographic analysis using substantially the same "rules of the game" as with limestones (e.g., Davis 1966).

The replacement of dolomite rhombs by calcite mosaic (dedolomitization) was observed by Folk in Ordovician carbonates (1959b), but the origin was not understood. Shearman, Khouri, and Taha (1961) reviewed early literature on dedolomitization (mainly Russian) that proposed flushing of the dolomite-bearing rock with high-Ca waters, either at the present weathering surface, under unconformities (Schmidt 1965), or in evaporite deposits that are being leached (Goldberg 1967; Evamy 1967).

Useful recent reviews on dolomite have been written by Fairbridge (1957), Friedman and Sanders (1967), and Michard (1969).

DIAGENESIS

Before the 1950's, little attention was paid to the diagenesis of carbonate rocks. Sorby (1879) had investigated the alteration of aragonitic skeletons to calcite and puzzled over criteria for distinguishing pore-filling from recrystallization spar. The magnificently detailed work of Cullis (1904) on the paragenetic sequence of cementation and recrystallization of aragonite, calcite, and dolomite was all too long ignored. He recognized fibrous crusts of aragonite and calcite, overgrowths on fossils, coarse pore-filling crystals of calcite and dolomite, and post-cementation internal sediment. The great majority of geologists thought, rather vaguely, that sparry calcite formed by some mysterious process of mud recrystallization; or was precipitated in some occult realm of deep burial. Cementation of beach-rock was well known, but the mode of lithification of most limestones was little investigated.

The carbonate diagenetic bandwagon was heralded by significant papers by Graf and Lamar (1950), Bergenback and Terriere (1953), and Bathurst (1958, 1959) who tried to develop criteria for deciphering the cementation vs. recrystallization origin of sparry calcite. Folk (1959*a*, 1959*b*) believed that nearly all sparry calcite was the result of pore-filling cementation, but made no attempt to understand the details of how this took place. Folk (1965*b*) and Wolf (1965*b*) developed complex classifications of spar types.

The big push in diagenesis came from the oil companies (particularly, Shell) who were interested in the creation and disappearance of precious pore space. In particular, R. J. Dunham became the archproponent of a fresh-water origin for most sparry calcite cement, implying that the cement was deposited from meteoric waters during subaerial exposure. This viewpoint was supported by many convincing studies, especially those on Recent or Pleistocene carbonates of lush tropical islands so easy and pleasant to study. Geologists tended to ignore the complex processes operating within the bowels of the earth or beneath the obscuring waters of the sea.

Inevitably the pendulum began to swing back. In the mid-1960's the sudden popularity of scuba diving, dynamiting of reefs, and the expanded collection of rocks from the deep-sea led to common discoveries of submarine cementation where there was no possibility of fresh-water flushing, and geologists began to obtain a more balanced picture of the many realms in which cementation can take place. It is quite probable that the most occult realm, that of subsurface cementation, will become the next great vogue as we learn how to sample and study this remote region. Floods of carbonate diagenesis papers threaten to inundate the nonspecialist; review papers have been written by Bathurst (1964), Chilingar, Bissell, and Wolf (1967), Purdy (1968), Friedman (1968), and Gavish and Friedman (1969); and more recently two substantial compilations: Fuchtbauer (1969) and Bricker (1971).

Diagenesis, like most other geologic processes, can be divided into a spectral range of types. A convenient division is as follows (Fig. 1):

(1) **Meteoric diagenesis**, generally, but not necessarily, caused by fresh water; divisible into (a) surficial—that taking place as crusts about streams or springs and in the soil zone as caliche; (b) vadose—above the water table; and (c) phreatic—extending from the water table to indeterminate (and often great) depths.

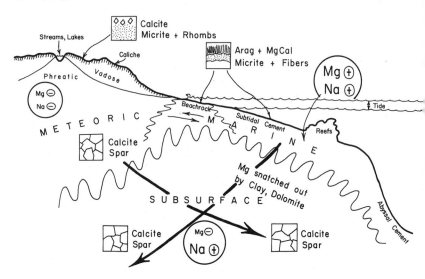

Fig. 1. Schematic diagram showing the relationship between the mineralogy and morphology of carbonate cements and their diagenetic environments.

In the intertidal and near-surface subtidal zones where both magnesium and sodium ions are relatively high, the cements are fibrous to micritic aragonite and magnesium calcite. In the zone of meteoric waters where sodium, magnesium, and the accompanying anions are relatively low, micritic calcite forms at the surface through rapid physico-chemical precipitation or as a result of metabolism of microorganisms. At depths within the zone of meteoric water, where crystallization is slow, calcite occurs as equant spar.

In the subsurface realm, where meteoric waters mix with saline and connate waters, the sodium ion concentration may range from very low to well above that of normal sea water. The Mg/Ca declines in this zone as a result of the removal of magnesium ion by clay minerals and the formation of dolomite. This reduction in magnesium ion removes the poisoning effect of this ion on the morphology of calcite, and sparry calcite is the characteristic cement. Aragonite and magnesium calcite invert to calcite, which forms pseudomorphs of the original minerals.

2) Peritidal diagenesis, including supratidal (areas subject to occasional flooding by sea water during highest storm tides) and intertidal (beachrock) realms. This is a zone where there are great fluctuations in salinity, from torrential rains that saturate an exposed beach or mudflat with fresh water to evaporation of sea water in residual hypersaline pools.

3) Subtidal diagenesis effected by nearly normal sea water in unconsolidated contemporary sediment grains that are still moving about, or as crusts from the surface to a few meters down in the sediment. Cementation has been recorded from a wide depth range, from the low tide mark to abyssal oceanic depths (Bricker, 1971).

4) Subsurface diagnesis, caused by sea water or meteoric water whose composition is modified significantly by passage through a large thickness of intervening sediments. This zone is the most difficult to define precisely and it passes without sharp break into (1), (2), and (3) above and into the hydrothermal zone.

A major result of our studies of diagnesis is the recognition of a great diagenetic schism between (1) cementation in sea water, which is always (?) aragonite or magnesian calcite, and always (?) finely fibrous or micritic; and (2) cementation in fresh or brackish water which is generally calcite and usually occurs as fairly large, equant mosaics of crystals. Ancient limestones are invariably cemented by calcite, which occurs in micritic, fibrous, and equant forms; but whether these cements are primary precipitates, or whether they result from the transformation of earlier minerals and morphologies into sparry calcite, is still a major mystery (Figs. 1, 2).

Meteoric Cementation

Extensive cementation occurs through the attack of meteoric water, usually fresh but not necessarily so, and at depths from the ground surface (caliche) to deep into the subsurface—as in the fresh-water lenses that extend far below sea level in islands and along coasts.

Dana (1849), Sorby (1862, 1879), Skeats (1903), David and Sweet (1904), and Field (1919) realized the potent effect of rain water in dissolving aragonitic fossils and reprecipitating the material elsewhere as calcite cement. In the Bahamas, Illing (1954) found carbonate rock above sea level was cemented with calcite; Ginsburg (1957) made similar observations in Florida and pointed out the lack of lithification of submarine sediments. Dunham was the major proponent of this idea in the late 1950's, though his ideas were not published until later (Dunham 1963, 1969*a*, 1969*b*).

An outstanding paper that pushed the scales far to the "meteoric diagenesis" side was that of Schlanger (1963; also Ladd & Schlanger 1960) who showed in borings on Eniwetok that immediately beneath unconformities aragonite fossils had dissolved and the pores were filled with sparry calcite.

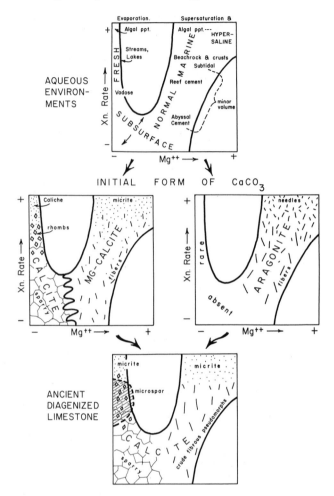

Fig. 2. Diagrams showing the relationships between carbonate morphology, magnesium ion concentration, and rate of crystallization.

The uppermost diagram illustrates common formative waters. At high crystallization rates (e.g., surface waters) there is a distinct gap between fresh waters, very low in magnesium ions, and sea water, high in magnesium. In the subsurface, with slower crystallization rates, there is a continuous range of magnesium ion concentrations from that of sea water to almost nil. In most subsurface waters Mg/Ca is under half that of sea water, owing to capture of magnesium by clays and dolomite.

The middle two diagrams show the variations in mineralogy and morphology. In waters with high concentrations of magnesium ion, fibrous (or micritic) Mg-calcite and aragonite form; in waters with low concentrations of magnesium ion, micritic or sparry calcite forms. Calcite spar forms in the meteoric to subsurface realm, in the latter through the selective loss of magnesium ion from carbonate waters.

The lowermost diagram shows that in ancient limestones, all these minerals are converted to calcite, which retains the grosser aspects of original morphology.

Calcite tends to form small rhombs in low magnesium (and low sodium) waters; it occurs either as a primary precipitate (in some caliches and fresh water carbonates), or by recrystallization to subrhombic microspar in ancient limestones when the imprisoning cage of magnesium ions is flushed from lithified micrite.

Outside the zone affected by meteoric water the aragonitic fossils survived and there was no calcite cement. This was repeated in several cycles. Logan (1969) showed that in Yucatan shelf sediments exposed by Pleistocene sea-level regression were cemented by sparry calcite, whereas those continually bathed in sea water were uncemented; and Illing, Wood, and Fuller (1967) described geosynclinal carbonates in the Middle East that were uncemented in areas of continuous subsidence and permanent sea-water saturation, while those in areas of shallow shelves, affected by folding and unconformities, were spar-cemented. Thus support grew for the idea that much, perhaps the predominant amount, of equant mosaic sparry calcite in limestones was formed either by subaerial exposure or flushing by meteoric water. Friedman (1964, 1968), Mathews (1967), and Land, McKenzie, and Gould (1967) reviewed the fresh-water diagenetic picture. Gross (1964) showed by analysis that sparry calcite had low O^{16}/O^{18} and C^{12}/C^{13} consistent with a fresh-water origin.

Fresh-water flushing not only changes the ratios of carbon and oxygen isotopes but expels magnesium from magnesian calcite (generally without any visible microscopic change in texture), and flushes out trace elements such as strontium. The change may be very rapid, in hundreds to thousands of years, and the rate of alteration also depends on rainfall (Mathews 1968).

Some calcite precipitates as tiny rhombohedra in low-Na^+ fresh-water environments with very low Mg/Ca ratios such as that found in some vadose-cemented eolianites, caliche, and stream deposits (Folk, 1971) and tufa (Irion & Müller 1968). Vadose carbonate can be expected to form mini-stalactites on the bottoms of grains (Dunham 1971), but the same morphology can occur in beachrock above the water level (Taylor & Illing 1969). Land (1970) proposes that phreatic diagenesis is far more rapid than vadose diagenesis.

Intertidal Cement (Beachrock)

The first form of diagenetic carbonate to be studied was beachrock formed on tropical coasts, generally on carbonate beaches but sometimes on beaches of silicate sand. Captain Robert Moresby (1835) had noted that it was harvested by natives for construction, and would harden upon exposure to air a few days after cutting into blocks. Darwin (1842) noted the "husk" of $CaCO_3$ deposited around the grains and also realized its rapid rate of formation. J. D. Dana (1849, pp. 44, 153; 1851, p. 368) proposed the mode of origin that is most popular today: beachrock was caused mainly by evaporation of sea water in the intertidal zone. Moresby (1835) and Branner (1904) thought it was formed at the zone of mixing between fresh and sea water, and David and Sweet (1904) favored an origin by percolating rain water, as did Russell (1962). It is now obvious that coastal carbonate rock can form in several ways. Beachrock cemented by aragonite or magnesian calcite (generally as fibrous crusts, sometimes as lumpy micritic coatings) is generally formed by evaporating sea water (Ginsburg 1953; Taylor & Illing 1969), sometimes influenced by organisms (Kaye 1959); while the equant crystals of sparry calcite form mainly at higher topographic levels by freshwater attack. Origin and properties of this type of cementation have been reviewed by Stoddart and Cann (1965) and Wolf (1965*b*).

Submarine Cement

Purdy (1968) has recently reviewed very early diagenetic alteration of grains. We consider here only pore-filling cementation in the subtidal zone.

Most early workers, if they thought about it at all, presumably guessed that limestones became cemented after burial in sea water or its connate modification, but there was no evidence for this view. Then, in the late 1950's, the spotlight suddenly shifted to a subaerial origin for mosaic calcite spar. The inevitable counterreformation set in, and now submarine cementation is the latest of the many diagenetic bandwagons to attract grantologists.

Friedman (1964, p. 806) mentioned lithified micrite at 150 fathoms. Next Milliman (1966) described in detail fibrous Mg-calcite crusts and lithification of carbonate sediment on seamounts, owing to a very slow rate of deposition; and Gevirtz and Friedman (1966) reported fibrous aragonite crusts in the saline pools at the bottom of the Red Sea. Within a couple years, many more localities had been discovered (Fischer & Garrison 1967, deep-sea sediments; Ginsburg *et al.* 1967, and Land & Goreau 1970, in dynamited reefs). Shinn (1969) discovered large areas of the Persian (Arabian) Gulf covered with a lithified hardground and proposed that the critical factor favoring cementation was a slow rate of deposition. At one stroke, this idea is immediately applicable to the supposed shallow-water Paleozoic spar-cemented limestones; it has been applied to Cretaceous rocks by Rose (1970).

So far, there appears to be no obvious difference in morphology or mineralogy between cement in beachrock, in shallow marine reefs and hardgrounds, and in deep-sea cements. In all these localities the cement may be either aragonite or magnesian calcite; the most common occurrence is as a radial fibrous crust or overgrowth, but micritic coats are also commonly found. Cements may fill voids such as burrows or hollow shells, may occur as intergranular cement, or as a lithification of micrite. In both beachrock and submarine cements organisms are thought to play a significant role.

Whether a sediment becomes cemented or not obviously depends upon the interaction of two factors: (1) rate of cement precipitation as influenced by evaporation, temperature, agitation, and organic influences; and (2) bottom sediment stability as influenced by current and wave disturbance, bioturbation, etc. As a generalization, precipitation is more rapid in the intertidal zone than in the subtidal zone; therefore, the proportion of carbonate beach sediments cemented to form beachrock is greater than the proportion of submarine sediments that are cemented to form "subtidal-rock." However, the volume of subtidal sediment so far exceeds the volume of beaches that subtidal rock is today probably far more abundant than beachrock.

Subsurface Cement

Little is known about this most remote realm of cementation. Purdy (1968) has reviewed some of the factors that may be effective. In my opinion subsurface cement will eventually be recognized as the most abundant type in ancient limestones. In the following section I suggest that subsurface cement is mainly rather coarsely crystalline sparry calcite mosaic and that it has this form because it crystallized from deeply buried, magnesium-deficient, connate waters.

Hypothesis on Carbonate Mineralogy and Morphology as Controlled by Magnesium Content and Rate of Precipitation

In reviewing the subject of diagenesis, the reader tends to get bewildered by the enormous amount of often-repetitive literature, the plethora of environments and micro-environments, and the variety of crystal sizes, shapes, and compositions involved in the diagenetic process. However, I think it is possible to reduce all this complexity to a simple and logical pattern which will explain most occurrences. Thus, the following diagenetic model is proposed under the philosophy of "Damn the variables, Full Speed Ahead!"

The four main rock-forming carbonate minerals are aragonite, calcite, magnesian-calcite, and dolomite. These can exist in three main crystal morphologies: micrite, fibers, and coarser euhedral to anhedral, equant crystals.

Micrite includes all crystals of maximum crystal dimension less than 4 microns. Electron microscopic study reveals that some micrite is in the form of tiny needles, some occurs as unit rhombohedra, and some may occur as steep-sided rhombs or scalenohedra. By definition all micrite has three small dimensions and it nearly always looks like fine, semi-opaque "mud" in the light microscope, no matter whether it exists as a bulk rock component, or as a thin crust on allochems. **Fibrous** carbonate includes crystals with one large and two small dimensions. Typically the width of fibers varies from a fraction of a micron up to perhaps several microns, and the fibers range from 6 (arbitrary cutoff) to 100 or more times as long as they are wide. **Sparry** calcite and dolomite form subequant crystals with three large dimensions, and may take the form of perfect rhombohedra, subhedral crystals, or apparently anhedral mosaics. As in any natural system, all gradations exist between these three end-points (micrite, fibers, and euhedral to anhedral spar) and boundaries and definitions are arbitrary.

Two major factors believed to explain the different carbonate minerals and different morphologies are (1) the rate of crystallization and (2) the effect of magnesium and other ions in the precipitating waters (Fig. 1).

Micrite can consist of aragonite, calcite, magnesian-calcite, or dolomite; all can form under a wide range of conditions. Micritic dolomite can form by direct precipitation or very early replacement in hypersaline marls or supratidal crusts, or in caliches. The several types of micritic calcium carbonate can form by organic precipitation as skeletal components (e.g., as invertebrate skeletons or as aragonite needles in green algae), as byproducts or photosynthesis (e.g., bacteria and blue-green algae), as possible direct chemical precipitates (needles and "whitings" in Persian Gulf muds), as micritic cements in beachrock or submarine sediments, and as caliche crusts in soils. In all these instances the micritic texture is probably the result of very rapid precipitation, either by organic influence or by evaporation or chemical reactions causing rapid supersaturation and multiple nucleation.

Fibrous texture is more complex in origin. Dolomite and calcite are seldom if ever finely fibrous. Magnesian calcite and aragonite are nearly always fibrous, and at present it appears that all submarine or beachrock cements are fibrous or micritic. As explained later, the role of dissolved ions in sea water (particularly magnesium) is believed to account for the lack of development of large, equant crystals of calcium carbonate in the marine environment, a slower rate of crystallization or fewer nuclei promotes the fibrous rather than the micritic morphology.

Coarser, equant crystals occur in dolomite and calcite, but not in aragonite or magnesian calcite. No such spar is known to be forming in contemporary marine sediments, such as submarine reefs, beachrocks, or shallow marine sediments. Coarser dolomite is the product of later replacement by solutions moving through buried or lithified sediments. Coarser sparry calcite forms in fresh-water-influenced sediments, or in deeply buried or ancient limestones where the formative process is not clear. The coarseness presumably indicates a slow rate of precipitation, probably from waters low in organic substances as well as magnesium and other ions.

Influence of Magnesium and Other Ions on Morphology of Carbonates

Leitmeier (1910–15) showed experimentally that the precipitation of aragonite from solution is promoted by a high magnesium ion concentration.

He also found that aragonite would only form in magnesium-ion-rich environments (including the vicinity of basic igneous rocks). A half-century later this work was continued (e.g., Lippman 1960; Taft 1963; Berner 1966, 1967; Bischoff 1968; Bischoff & Fyfe 1968). Thus, in sea water rich in magnesium ion (about 0.13 percent), only magnesian calcite or aragonite crystallize, and calcite is not known to form inorganically in any quantity. Furthermore, in sea water only micrite or fibers develop, not coarser sparry crystals.

The works cited above propose that magnesium ions prevent ordinary calcite from forming by "poisoning" the growing crystal's faces, blocking further crystallization. Consequently, if Mg/Ca is over about 2:1, the $CaCO_3$ goes mainly into aragonite because of the difficulty of forming calcite (Lippman 1960). Such calcite lattices as are able to develop must absorb magnesium ions into their structure, becoming magnesian-calcite, and the struggling minute crystals are forced into a fibrous habit. No explanation has yet been offered for the fibrous nature. I propose that *selective* poisoning of all the crystal growth directions but one—the *c* axis direction—forces magnesian calcite to assume the fibrous form, using the following chain of reason (Fig. 3):

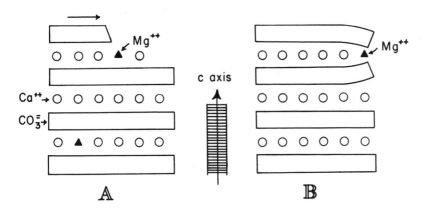

Fig. 3. Diagram illustrating the proposed explanation for the preferential growth of magnesium calcite as fibers parallel to *c* axis. The *c* axis is perpendicular to alternating sheets of carbonate and calcium ions. If a magnesium ion is added to the end of a growing crystal as in A, it can easily be buried by the next overlapping carbonate sheet without distorting the structure. However, if the small magnesium ion is added to the side of the crystal, as in B, the neighboring carbonate sheets close together to accommodate it, hampering further sideward growth. Therefore, crystal growth is much more rapid in the *c* axis direction, and sideward growth is blocked by selective magnesium poisoning. This effect means (1) it is difficult for any magnesian calcite to form, and calcium carbonate crystallizes much more readily as aragonite; (2) any magnesian calcite that does form will be very small crystals; and (3) all such tiny crystals will be fibrous or have steeply inclined crystal faces.

1. In calcite crystals, Vater (1899) has shown that increasing the content of sulphate ion in solution makes the crystals change from the simple, unit rhomb face to rhombs or scalenohedra of ever more steep slope, until finally a crystal with prism faces results.

2. Leitmeier (1910, 1915) showed that increasing amounts of magnesium ion made calcite form steeper rhombs, scalenohedra, and prism faces. At very low Mg/Ca only the unit rhomb developed.

3. Bischoff and Fyfe (1968) show that *both* sulphate and magnesium ions inhibit growth of calcite crystals.

4. If both sulphate and magnesium ions promote the growth of crystals parallel to the *c* axis, then I propose that both have the selective effect of "poisoning" and suppressing the growth of the *side* faces.

5. In metals, the presence of impurities also promotes the growth of "whisker" crystals (Price, Vermilyea, & Webb 1958; Brenner 1963). Foreign atoms poison the growth of side faces by stopping growth of the successive layers of the crystal lattice, while the tips of the "whiskers" either remain clean or bury the foreign ions too rapidly for them to have an effect.

We still must explain why it is that magnesium and sulphate ions poison only the side faces of calcite and allow the crystals to grow in the *c*-axis direction. The following explanation is proposed: The calcite lattice consists of alternating layers of carbonate groups and calcium ions (Fig. 3). Imagine the "c" end of a growing calcite crystal (i.e., the basal pinacoid). A sheet of calcium ions has been added to the crystal, and it is being overlapped by another sheet of carbonate groups. If a magnesium ion lands in the calcium sheet, it will easily be sidled up to by more calcium ions and covered over by the next carbonate sheet, no harm will be done, and the crystal will continue to grow out in the *c*-axis direction, burying its mistakes.

But now imagine the side of such a crystal with exposed alternating edges of calcium sheets and carbonate sheets. If a stray magnesium ion attaches itself to the edge of a calcium sheet, it will block growth in that direction (it cannot be covered over by a carbonate sheet), and being of smaller ionic radius it will make the over- and underlying carbonate sheets scrunch together to accomodate it. Thus contracted, the structure can no longer accept the next large calcium ion that comes along searching for its rightful place, and sidewise growth is slowed or stopped.

This distortion can explain why inorganic magnesian calcite only occurs as micrite crystals only a few microns wide (often with curved faces or small fibers) and never forms large crystals; only so much distortion can be tolerated before growth completely stops. It is now evident that the *lack* of Mg-poisoning allows the growth of large, sparry crystals of low-Mg calcite. This theory does *not* explain why aragonite is typically fibrous. Johnston and Williamson (1916) found that aragonite typically contains small amounts of SO_4, Pb, Sr, or Zn, and Schroeder (1969) also notes Mg. Perhaps the presence of these foreign ions produces the fibrous form by a similar selective side-poisoning mechanism.

Bunn (1933) proposed a neat mechanism to explain this phenomenon, and although he did not work with natural minerals, the same principle applies. Consider the example of aragonite and another carbonate of similar structure but with a different cation, such as strontium. If one looks down the *c*-axis direction, both of these minerals have nearly the same crystal-chemical appearance, as the atomic lattice widths are controlled mainly by the dimensions of the carbonate sheets, which are fairly similar in the case of these two minerals. However, looked at from the side, the different sizes of the metal cations (Ca vs. Sr) means that the carbonate sheets will have a greater difference in spacing. Consequently, if contaminating Sr ions are present in a solution crystallizing aragonite, they will distort the *c*-direction spacing of the carbonate sheets; thus growth continues in the *c*-axis direction but sideward growth is retarded by accumulating misfits, hence the mineral becomes fibrous parallel to *c*. I wish to apply this explanation to both magnesian calcite and aragonite.

Dolomite assumes no fibrous form, showing a complete gradation (almost uniform) in crystal size from less than one micron up to several mm. Why? In the dolomite crystal, layers of carbonate groups are separated by alternate layers of calcium and magnesium ions. Thus any stray magnesium ions coming in from the side can be incorporated in the nearest proper layer without harmful effect. Lateral growth is not impeded, and growth of equant, rhombic crystals proceeds.

Geological Explanation of Morphology: Realms of Magnesium and Sodium Ions

Sea water produces micritic or fibrous aragonite or magnesian-calcite. Fresh water produces micritic or sparry calcite. I propose that deep subsurface waters usually produce sparry calcite. Why these relationships (summarized in fig. 2)?

In sea water, magnesium ions are easily incorporated in the calcite lattice, so that any calcite that does form will be rich in magnesium. However, *side* faces of the growing calcite crystals are selectively poisoned so that only short fibers or micrite develop, and no large equant crystals can form. Cements of magnesium calcite typically show steep rhomb faces indicative of magnesium-poisoning of sideward growth.

Fresh water contains very little magnesium (1–10 ppm). Thus ordinary calcite forms, either micritic (if precipitated upon total evaporation or by photosynthesis, etc. as in soil crusts and caliches), or as equant, sparry crystals, often as good rhombohedra where precipitated by a continuous supply of fresh water (as in nodular caliches, fresh-water ponds, streams, algal tufa, vadose zones, etc.). *Finely* fibrous fresh-water calcite is rare, but could presumably form by an intermediate rate of crystallization.

I believe that much of the coarse-grained, calcite spar mosaic that is the common cement of many carbonate rocks and sandstones must have crystallized from deeply buried connate waters low in magnesium. I propose that magnesium is removed selectively from connate waters at shallow depths either by clay minerals, chlorite or montmorillonite, or by the growth of replacement dolomite. The change in Na/Mg from 9:1 in sea water to 18:1 in subsurface brines points to capture of about half of the magnesium by dolomite or clays (Degens 1965, p. 189; Graf et al. 1966). Mg/Ca is reduced from approximately 3:1 in sea water to 1.5:1 or less in brines, which is apparently low enough so that magnesium-poisoning no longer forces a fibrous morphology on calcite.

The typical paragenetic sequence in ancient limestones of fibrous calcite crust following equant mosaic calcite can now be explained in two ways: (1) Initial beachrock or submarine cementation, followed by passage of the sediment into fresh water by relative fall of sea level, or by subpercolation of fresh water (Fig. 1) in aquifers far out under marine-shelf sediments (e.g., Manheim 1967). Sparry calcite is here precipitated by magnesium-free meteoric water. (2) Initial beachrock or surficial submarine cement, followed by fairly deep burial in connate or saline brines derived from sea water from which much of the magnesium has been taken by clays or by dolomitization. Blocky sparry calcite is here precipitated by magnesium-poor but sodium-rich sea water. Mingling of meteoric with connate water will further lower the magnesium concentration.

Magnesium, Microspar, and the Micrite Curtain

Aspects of recrystallization have been covered elsewhere (Bathurst 1958, 1959, 1964; Folk 1959*a*, 1965*b*; Mišík 1968) and will not be reviewed. Neomorphism of micrite to microspar is volumetrically the most important form of recrystallization; the cause of this change has been elusive, but the following new revelation is presented as a hypothesis.

In the *normal marine* environment, the abundance of magnesium ion means that either magnesian-calcite or aragonite will form initially, as minute needles or as fibers, because of the selective poisoning effect. Ordinary calcite does not form in this environment.

Upon burial and advanced diagenesis of carbonate mud, the aragonite needles invert to calcite, the magnesian calcite expels the magnesium ion, and new material is precipitated to fill up the original great microporosity. The result is a uniform mosaic of polyhedral, equant blocks of calcite averaging about 2 microns. The magnesium ions expelled from the new, tiny calcite crystals are retained interstitially in the rock, probably attached to the surfaces of calcite polyhedra; they form a sort of "cage" around each calcite crystal and their distortion of the lattice prevents normal growth beyond the equilibrium size of 2-3 microns, which also represents the average *length* of the original needles in the carbonate ooze. Hence, the worldwide uniformity of crystal size in micrite (Folk 1965*b*); it is a chemically induced limitation caused by small amounts of interstitial magnesium ion.

If for any reason the surroundings become very low in magnesium, the calcite crystals are freed of their imprisoning cage of magnesium ions; they can burst through the "micrite curtain" and by porphyroid neomorphism transform to microspar or even pseudo-spar. The modal size of microspar is about 5 microns, representing a doubling of micrite grain diameter, or an eight-fold increase in grain volume and an eight-fold decrease in grain surface area. The favored crystal size of calcite in a magnesium-free environment is apparently about 5 microns. These two equilibrium crystal sizes in respectively magnesium-rich and magnesium-poor environments explain the gap in micrite-microspar crystal sizes at about 3-4 microns (Folk 1965*b*).

Magnesium can be removed from the surroundings in several ways:

1) Deposition of the original sediment in a brackish environment with initially low magnesium; the brackish water may serve to remove even more of the magnesium by leaching (Berner 1966).

2) Percolation of fresh waters into carbonate strata below; Carozzi (1971) proposed this mechanism to explain how limestones underlying Pennsylvanian fluvial channels have been converted to microspar, but those away from the percolating fresh waters remain normal micrite.

3) Seizure of magnesium ions by interbedded or inter-carbonate-crystal clay minerals as waters percolate through the rock; clays tend to work toward a montmorillonite or chlorite composition by capture of magnesium.

4) Flushing of magnesium by meteoric waters in weathering. This type of late diagenesis is very important in recrystallization of some limestones as well as in dedolomitization; the two phenomena often occur together.

The above hypothesis explains the long-known observation that microspar is associated with brackish or clayey carbonates (Folk 1962*a*, 1965*a*) and with outcrop weathering. Choquette (1968) showed that microspar had isotopic compositions related to fresh water, while micrite in unaltered limestones had normal marine ratios.

In the *fresh-water* environment, little magnesium ion is present. Calcite crystals thus have no magnesium-induced upper size limit. Cave muds, for example, show spectacular diagenesis of calcite (Frank 1965). Of course the super-rapid, biochemical or evaporation-to-dryness precipitates of calcite (as in fresh-water algal crusts or caliches) are very fine-grained. But other calcite that encrusts stream plants, or that forms on the surfaces of streams has no ion-limitation on size although 5-10 microns seems to be the most common size.

The maximum size for rhombic fresh-water calcite also appears to be about 5-10 microns. Very low concentrations of sodium appear to favor rhomb development (Figs. 4,5). Above 5-10 microns calcite becomes more anhedral, just as dolomite is most euhedral at about 100-300 microns and becomes more anhedral above that size. Euhedral to subhedral calcite rhombs of about 5-10 microns diameter are common in some nodular caliches, they crystallize on the surface of some creeks and as algal crusts in streams, they occur as vadose cements in some eolian carbonates, examples in Perkins (1968), and appear to make up Dunham's (1969*a*) vadose silt. Strangely enough this 5-10 micron size is the typical occurrence of microspar which, because it is recrystallizing in a solid, unyielding mass only forms imperfect, loafish rhombs. Also, a high concentration on foreign ions may prevent rhomb formation (determined in lab experiments by Nickl and Henisch 1969); thus no rhombs form in the highly saline, but magnesium-poor, subsurface brines. Instead there occur complex-faced spar crystals that look anhedral in thin section.

The subject of carbonate diagenesis can be summarized by the simple diagram of Figure 5.

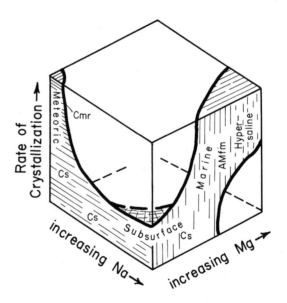

Fig. 4. Schematic block diagram summarizing the relationship between the morphology of carbonate cements, the concentration of sodium and magnesium ions, and rate of crystallization. M-magnesium calcite; A-aragonite; c-calcite, m-micritic; r-rhombic; f-fibrous; s-coarse equant spar.

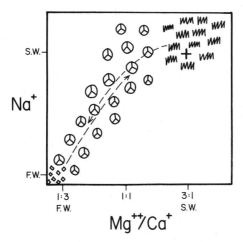

Fig. 5. Diagram showing the relationship between the morphology and mineralogy of inorganically precipitated carbonates, the Mg^{++}/Ca^{++} ratio, and the concentration of sodium ions. It is assumed that the concentration of chloride, sulphate, and other anions is proportional to those of sodium ions.

The fibrous symbols, upper right, are restricted to waters of normal or slightly elevated salinity. Fibrous aragonite or Mg-calcite form first, and are then changed to fibrous calcite as magnesium ion is removed during burial.

Circular peace symbols lie in the realm of connate brines to subsurface meteoric waters, where mixing is indicated by the arrows; here equant, blocky, sparry calcite forms.

Tiny rhomb symbols lie in the realm of nearly pure fresh water. Small calcite crystals tend toward a simple rhomb form; larger crystals become blocky and more complex.

Conclusions

What then have we learned about carbonates in the post-Sorbian Age? First, carbonates are complex and fascinating beyond anyone's dreams in the 1940's. The number of man-hours spent on carbonate petrology has increased at an astronomical rate, exemplified when nearly fifty geologists, all of whom are working intensively on carbonate diagenesis, can gather at a cementation conference. The volume of research on carbonate rocks alone probably outweighs research done on igneous and metamorphic rocks combined.

The major advances have been in the understanding of diagenesis and in interpretation of depositional environments. Earlier geologists could say little more than "shallow" or "deep" and at that be guessing. In standard oil company logs or stratigraphic sections a description such as "gray, crystalline limestone" would have been considered quite detailed. We know now how to interpret detailed sequences of carbonate mini-environments, and how they are integrated into a stratigraphic and depositional framework based upon the geographic arrangements we see today in intensively researched areas like the Bahamas, Florida, the Arabian coast, or Yucatan-Honduras. Studies of trace elements, isotopes, luminescence petrography, and electron microscopy can contribute little gems here and there to the crown, but the main interpretive contributions are still made by careful field description and petrographic work.

What can we say about the future of carbonate petrography? Past history really gives no very clear answer if one is looking for general principles. Some advances have been made by lone individuals, others have been made by well-financed teams. Some discoveries have come about by complete accident (most of the really *new* discoveries are made this way!) while others have been made through painstaking slogging after a well-defined and planned goal.

Speaking personally, most of my own "discoveries" have arisen by accident. The rhombic theory of fresh-water calcite came about because of a chance observation by high-powered pocket microscope of a "pollution" scum floating on a creek near Austin; behold, it was composed of euhedral calcite rhombs. The theory on the side-poisoning effect of magnesium ions was born through reading in the metallurgical literature as to what made chalcedony fibrous, from there proceeding to factors that favored fibrous growth in general.

One can surely say that, considering everything from gross field relationships to thin section petrography to the intricacies of isotopes or the electron microscope, all are vital; no minute fact or curious "idle" observation is useless or wasted. One keeps filing away odd bits of information in the dusty corners of the mind until one day they all reach critical mass and produce a flash of inspiration. As Szent-Gyorgi has said: "Research is to see what everyone else has seen and to think what no one else has thought."

Acknowledgments. It is a pleasure to acknowledge Lynton Land, who has frequently acted as an undocile sounding board for ideas across the hall. I have also utilized class cadenzas prepared by many students, but in particular M. B. Cooper Waitt, G. Heiken, P. Scholle, J. Miller, R. Crawley, R. Laudon, G. Kessler, and R. Keir.

REFERENCES

Adams, John Emery, & Rhodes, Mary Louise. 1960. Dolomitization by seepage refluxion. *Am. Assoc. Petroleum Geologists Bull. 44*, 1912–20.

Alderman, A. R., & Skinner, H. Catherine W. 1957. Dolomite sedimentation in the southeast of South Australia. *Am. Jour. Sci. 255*, 561–67.

Atwood, D. K., & Bubb, J. N. 1970. Distribution of dolomite in a tidal flat environment, Sugarloaf Key, Florida. *Jour. Geology 78*, 499–505.

Ball, M. M. 1967. Carbonate sand bodies of Florida and the Bahamas. *Jour. Sedimentary Petrology 37*, 556–91.

Bathurst, R. G. C. 1958. Diagenetic fabrics in some British Dinantian limestones. *Liverpool and Manchester Geol. Jour. 2*, 11–36.

———. 1959. Diagenesis in Mississippian calcilutites and pseudobreccias. *Jour. Sedimentary Petrology 29*, 365–76.

———. 1964. Diagenesis and paleoecology: a survey, *in* Imbrie, J., & Newell, N. (eds.) *Approaches to paleoecology*, New York: John Wiley, pp. 319–44.

Bavendamm, Werner. 1932. Die Mikrobiologische Kalkfällung in der tropischen See. *Archiv fur Mikrobiologie 3*, 205–76.

———. 1931. The possible role of micro-organisms in the precipitation of calcium carbonate in tropical seas. *Science 73*, 597–98.

Beales, F. W. 1958. Ancient sediments of Bahaman type. *Am. Assoc. Petroleum Geologists Bull. 42*, 1845–80.

———. 1961. Modern sediment studies and ancient carbonate environments. *Jour. Alberta Soc. of Petr. Geol. 9*, 319–30.

———. 1965. Diagenesis in pelleted limestone, *in* Pray, Lloyd C. & Murray, Raymond C. (eds.), *Dolomitization and limestone diagenesis. Soc. Econ. Paleontologists and Mineralogists, Sp. Publ. 13*, pp. 49–70.

Behrens, E. William, & Land, Lynton S. 1970. Subtidal Holocene dolomite in Baffin Bay, Texas. *Geol. Soc. America Spec. Publ. 2*, no. 7, 491–92.

Bergenback, R. E., & Terriere, R. T. 1953. Petrography and petrology of Scurry Reef, Scurry County, Texas. *Am. Assoc. Petroleum Geologists Bull. 37*, 1014–29.

Berner, R. A. 1965. Dolomitization of the mid-Pacific atolls. *Science 147*, 1297–99.

———. 1966a. Chemical diagenesis of some modern carbonate sediments. *Am. Jour. Sci. 264*, 1–36.

———. 1966b. Diagenesis of carbonate sediments: interaction of Mg^{++} in sea water with mineral grains. *Science 153*, 188–91.

———. 1967. Comparative dissolution characteristic of carbonate minerals in the presence and absence of aqueous magnesium ion. *Am. Jour. Sci. 265*, 45–70.

Bischoff, J. L. 1968. Kinetics of calcite nucleation: magnesium ion inhibition and ionic strength catalysis. *Jour. Geophys. Research 73*, 3315–22.

Bischoff, J. L., & Fyfe, W. S. 1968. Catalysis, inhibition, and the calcite-aragonite problem. 1. The aragonite-calcite transformation. *Am. Jour. Sci. 266*, 65–79.

Black, Maurice. 1933. The precipitation of calcium carbonate on the Great Bahama Bank. *Geol. Mag. 70*, 455–66.

Bøggild, O. B. 1930. The shell structure of the mollusks. *D. Kgl. Danske Vidensk. Selsk. Skrifter, Naturvidensk. Og. Mathem. Afd. 9, Roekke 2*, 231–326.

Bramkamp, R. A., & Powers, R. W. 1958. Classification of Arabian carbonate rocks. *Geol. Soc. America Bull. 69*, 1305–18.

Branner, John Cooper. 1904. The stone reefs of Brazil, their geological and geographical relations, with a chapter on Coral Reefs. *Harvard Coll. Mus. Comp. Zool. Bull. 44*, 284 pp.

Brenner, S. S. 1963. [Vapor growth in] Metals, *in* Gilman, J. J., *The art and science of growing crystals.* New York: John Wiley & Sons, pp. 30–54.

Bricker, Owen P. (ed.). 1971. *Carbonate Cements,* The Johns Hopkins University Studies in Geology No. 19. Baltimore, Maryland: The Johns Hopkins Press, 358 pp.

Brown, Thomas C. 1913. Notes on the origin of certain Paleozoic sediments, illustrated by the Cambrian and Ordovician rocks of Centre County, Pennsylvania. *Jour. Geology 21*, 232–50.

———. 1914. Origin of oolites and the oolitic texture in rocks. *Geol. Soc. America Bull. 25*, 745–80.

Bucher, W. H. 1918. On oolites and spherulites. *Jour. Geology 26*, 593–609.

Bunn, C. W. 1933. Adsorption, oriented overgrowth and mixed crystal formation. *Proc. Roy. Soc. London, 141A*, 567–93.

Butler, Godfrey P. 1969. Modern evaporite deposition and geochemistry of coexisting brines, the Sabkha, Trucial Coast, Arabian Gulf. *Jour. Sedimentary Petrology 39*, 70–89.

Carozzi, Albert V. 1960. *Microscopic sedimentary petrology.* New York: J. Wiley, 485 pp.

———. 1971. *In* Bricker, Owen P. (ed.), *Carbonate Cements,* The Johns Hopkins Studies in Geology No. 19. Baltimore: The Johns Hopkins Press, pp. 205–6.

Carozzi, Albert V., & Lundwall, Walker R., Jr. 1959. Microfacies study of a middle Devonian bioherm, Columbus, Indiana. *Jour. Sedimentary Petrology 29*, 343–53.

Carpenter, William B. 1844, 1847. On the microscopic structures of shells. *Brit. Assoc. Adv. Sci. Rept. 14*, 1–24; *17*, 93–134.

Cayeux, Lucien. 1935. *Les Roches Sedimentaires de France, Roches Carbonatees.* Paris: Masson et Cie, 447 pp.

Chanda, S. K. 1967. Petrogenesis of the calcareous constituents of the Lameta Group around Jabalpur, M. P., India. *Jour. Sedimentary Petrology 37*, 425–37.

Chave, Keith E. 1954. Aspects of the biogeochemistry of magnesium, 1. Calcareous marine organisms. *Jour. Geology 62*, 266–83.

———. 1964. Skeletal durability and preservation, *in* Imbrie, J., & Newell, N. (eds.), *Approaches to paleoecology,* New York: John Wiley & Sons, pp. 377–87.

———. 1965. Carbonates: association with organic matter in surface seawater. *Science 148*, 1723–24.

Choquette, Philip W. 1968. Marine diagenesis of shallow marine limemud sediments: insights from δO^{18} and δC^{13} data. *Science 161*, 1130–32.

Cloud, Preston E. 1962. Environment of calcium carbonate deposition west of Andros Island, Bahamas. *U.S. Geol. Sur. Prof. Paper 350*, 138 pp.

Cullis, C. Gilbert. 1904. The mineralogical changes observed in the cores of the Funafuti borings. *The Atoll of Funafuti*, Sect. XIV *Royal Soc. London*, pp. 392–420.

Dana, James D. 1843. On the analogies between the modern igneous rocks and the so-called primary formations, and the metamorphic changes produced by heat in the associated sedimentary deposits. *Am. Jour. Sci. 45*, 104–29.

_____. 1849. *Geology: vol. X of United States Exploring Expedition* (1835–42, under Charles Wilkes). New York: G. P. Putnam, 756 pp.

_____. 1851. On coral reefs and islands. *Amer. Jour. Sci.*, series II, *11*, 357–72; *12*: 25–51, 165–86, 329–38.

Darwin, Charles. 1842. *Geological observations on the volcanic islands and parts of South America visited during the voyage of the H.M.S. Beagle.* London: Smith, Elder & Co., 2nd ed., 647 pp.

David, T. W. Edgeworth, & Sweet, G. 1904. The geology of Funafuti. *The atoll of Funafuti*, section V, *Royal Soc. London*, pp. 61–124.

Davis, Richard A., Jr. 1966a. Willow River dolomite: Ordovician analogue of modern algal stromatolite environment. *Jour. Geology 79*, 908–23.

_____. 1966b. Quiet water oolites from the Ordovician of Minnesota, *Jour. Sedimentary Petrology 36*, 813–18.

Deffeyes, K. S., Lucia, F. J., & Weyl, P. K. 1965. Dolomitization of Recent and Plio-Pleistocene sediments by marine evaporite waters on Bonaire, Netherlands Antilles, *in* Pray, Lloyd C., & Murray, Raymond C., (eds.), *Dolomitization and limestone diagenesis*, Soc. Econ. Paleontologists and Mineralogists, Spec. Publ. 13, pp. 71–88.

DeFord, Ronald K. 1946. Grain size in carbonate rock. *Am. Assoc. Petroleum Geologists Bull. 30*, 1921–27.

Degens, Egon T. 1965. *Geochemistry of sediments.* Englewood Cliffs, N.J.: Prentice–Hall, 342 pp.

Drew, G. H. 1914. On the precipitation of calcium carbonate in the sea by marine bacteria, and on the action of denitrifying bacteria in tropical and temperate seas. Carnegie Inst. Washington, Publ. No. 182, *Papers Tortugas Laboratory 5*, 9–45.

Dunham, Robert J. 1962. Classification of carbonate rocks according to depositional texture, p. 108–121 *in* Ham, William (ed.), *Classification of carbonate rocks.* Am. Assoc. Petroleum Geologists Memoir *1*, 108–21.

_____. 1963. Early vadose silt in Townsend Mound (reef), New Mexico (abs.). *Am. Assoc. Petroleum Geologists Bull. 47*, 356.

_____. 1965. Vadose pisolite in the Capitan reef (abs.). *Am. Assoc. Petroleum Geologists Bull. 49*, 338.

_____. 1969a. Early vadose silt in Townsend Mound (reef), New Mexico, *in* Friedman, Gerald M. (ed.), *Depositional environments in carbonate rocks.* Soc. Econ. Paleontologists and Mineralogists Spec. Publ. 14, pp. 139–81.

_____. 1969b. Vadose pisolite in the Capitan reef (Permian), New Mexico and Texas, *in* Friedman, Gerald M. (ed.), *Depositional environments in carbonate rocks*, Soc. Econ. Paleontologists Mineralogists Spec. Publ. *14*, 182–91.

Evamy, B. D. 1967. Dedolomitization and the development of rhombohedral pores in limestone. *Jour. Sedimentary Petrology 37*, 1204-15.

Fairbridge, Rhodes W. 1957. The dolomite question, *in* Le Blanc, Rufus J., and Breeding, Julia C., *Regional aspects of carbonate deposition*, Soc. Econ. Paleontologists and Mineralogists, *Special Publ. 5*, 125-78.

Field, Richard M. 1916. A preliminary paper on the origin and classification of the intraformational conglomerates and breccias. *Ottawa Naturalist 30*, 29-36, 47-52, 58-66.

———. 1919. Investigations regarding the calcium carbonate oozes at Tortugas, and the beach-rock at Loggerhead Key. *Carnegie Inst. Washington Yearbook 18*, 197-98.

———. 1931. Geology of the Bahamas. *Geol. Soc. America Bull. 42*, 759-84.

Fischer, Alfred G., & Garrison, Robert E. 1967. Carbonate lithification on the sea floor. *Jour. Geology 75*, 488-96.

Folk, Robert L. 1952. Petrography and petrology of the Lower Ordovician Beekmantown carbonate rocks in the vicinity of State College, Pennsylvania. Ph.D. dissertation, Penna. State College, 336 pp.

———. 1957. Practical petrographic classification of limestones (abs.). *Am. Assoc. Petroleum Geologists Ann. Meeting*, pp. 58-59.

———. 1959*a*. Practical petrographic classification of limestones. *Am. Assoc. Petroleum Geologists Bull. 43*, 1-38.

———. 1959*b*. Thin section examination of pre-Simpson Paleozoic rocks, *in* Barnes, V.E., et al., *Stratigraphy of the Pre-Simpson Paleozoic subsurface rocks of Texas and Southeast New Mexico*. Univ. of Texas, Bur. Econ. Geol. Publ. 5924, 95-130, 236-91, 727-801.

———. 1962*a*. Petrography and origin of the Silurian Rochester and McKenzie shales, Morgan County, West Virginia. *Jour. Sedimentary Petrology 32*, 539-78.

———. 1962*b*. Sorting in some carbonate beaches of Mexico, *Trans*. New York *Acad. Sci. Ser. II, 25*, 222-44.

———. 1962*c*. Spectral subdivision of limestone types, *in* Ham, W. (ed.), *Classification of carbonate rocks*, Am. Assoc. Petroleum Geologists Mem. 1, 62-84.

Folk, Robert L., & Robles, Rogelio. 1964. Carbonate sediments of Isla Perez, Alacran Reef Complex, Yucatan. *Jour. Geology 72*, 255-92.

Folk, Robert L. 1965*a*. Henry Clifton Sorby (1826-1908), the founder of petrography. *Jour. Geol. Education 13*, 43-47, 93.

———. 1965*b*. Some aspects of recrystallization in ancient limestones, *in* Pray, Lloyd C., & Murray, Raymond C. (eds.), *Dolomitization and limestone diagenesis*, Soc. Econ. Paleontologists and Mineralogists Spec. Publ. *13*, 14-48.

———. 1967. Carbonate sediments of Isla Mujeres, Quintana Roo, Mexico and vicinity. *New Orleans Geol. Soc. Guidebook Field Trip to Peninsula of Yucatan*, Geol. Soc. Amer. Annual Meeting, pp. 100-23.

———. 1971. Caliche nodule composed of calcite rhombs, *in* Bricker, Owen P. (ed.), *Carbonate cements*, Johns Hopkins Studies in Geology No. 19. Baltimore: The Johns Hopkins Press, pp. 167-68.

Folk, Robert L., & Cotera, Augustus S. 1971. Carbonate sand cays of Alacran Reef, Yucatan, Mexico, Sediments. *Atoll Research Bull. 137*, 35 pp.

Force, Lucy M. 1969. Calcium carbonate size distribution on the west Florida shelf and experimental studies on the microarchitectural control of skeletal breakdown. *Jour. Sedimentary Petrology 39*, 902–34.

Fosberg, F. Raymond, & Carroll, Dorothy. 1965. Terrestrial sediments and soils of the Northern Marshall Islands, *Atoll Research Bull. 113*, 156 pp.

Frank, Ruben M. 1965. Petrologic study of sediments from selected central Texas caves, M.A. thesis, Univ. Texas, 117 pp.

Freeman, Tom. 1962. Quiet water oolites from Laguna Madre, Texas. *Jour. Sedimentary Petrology 32*, 475–83.

Friedman, Gerald M. 1964. Early diagenesis and lithification in carbonate sediments, *Jour. Sedimentary Petrology 34*, 777–813.

Friedman, Gerald M., & Sanders, John E. 1967. Origin and occurrence of dolostones, *in* Chilinger, G., Bissell, H., and Fairbridge, R. (eds.), *Carbonate rocks, Developments in Sedimentology 9* (Amsterdam, Elsevier), pp. 267–348.

Friedman, Gerald M. 1968. The fabric of carbonate cement and matrix and its dependence on the salinity of water. *in* Müller, German, & Friedman, Gerald M. (eds.), *Recent developments in carbonate sedimentology in Central Europe*. New York: Springer–Verlag, pp. 11–20.

Friedman, Gerald (ed.). 1969. *Depositional environments in carbonate rocks.* Soc. Econ. Paleontologists and Mineralogists, Spec. Pub. 14, 198 pp.

Füchtbauer, Hans, & Goldschmidt, Bertha. 1965. Beziehungeun Zwischen Calciumgehalt und Bildungsbedingen der Dolomite. *Geol. Rundschau 55*, 29–40.

Füchtbauer, H. (ed.). 1969. *Lithification of carbonate sediments*, I. Special Issue, *Sedimentology 12*, 1–160.

Gee, H. 1932. Preliminary experiments in precipitation by removal of carbon dioxide under aseptic conditions. *Univ. Calif., Scripps Inst. of Oceanogr., Tech. Ser. Bull. 3*, 180–87.

Gevirtz, Joel L., & Friedman, Gerald M. 1966. Deep-sea carbonate sediments of the Red Sea and their implications on marine lithification. *Jour. Sedimentary Petrology 36*, 143–51.

Ginsburg, Robert N. 1953. Beachrock in South Florida, *Jour. Sedimentary Petrology 23*, 85–92.

———. 1956. Environmental relationships of grain size and constituent particles in some South Florida carbonate sediments. *Amer. Assoc. Petroleum Geologists Bull. 40*, 2384–2427.

———. 1957. Early diagenesis and lithification of shallow-water carbonate sediments in South Florida, *in* Le Blanc, Rufus J., and Breeding, Julia G. (eds.), *Regional aspects of carbonate deposition*, Soc. Econ. Paleontologists and Mineralogists Spec. Publ. 5, 80–99.

Ginsburg, Robert N., Shinn, E. A., & Schroeder, Johannes H. 1967. Submarine cementation and internal sedimentation within Bermuda reefs (abs.), *Geol. Soc. America Spec. Paper 115*, 78–79.

Goldberg, Moshe. 1967. Supratidal dolomitization and dedolomitization in Jurassic rocks of Hamakhtesh Hagatan, Israel. *Jour. Sedimentary Petrology 37*, 760–73.

Grabau, A. W. 1903. Paleozoic coral reefs. *Geol. Soc. America Bull. 14*, 337–52.

Graf, D. L., & Lamar, J. E. 1950. Petrology of Fredonia oolite in southern Illinois. *Am. Assoc. Petroleum Geologists Bull. 34*, 2318–36.

Graf, D. L., Meents, W. F., Friedman, I., & Shimp, F. F. 1966. The origin of saline formation waters, III: Calcium chloride waters. *Illinois Geol. Surv. Circ. 397*, 60 pp.

Gross, M. Grant. 1964. Variation in the O^{18}/O^{16} and C^{13}/C^{12} ratios of diagenetically altered limestones in the Bermuda Islands. *Jour. Geology 72*, 170–94.

Ham, William E. (ed.). 1962. *Classification of carbonate rocks. Am. Assoc. Petroleum Geologists Mem. 1*, 229 pp.

Hatch, F. H., Rastall, M. A. & Black, Maurice. 1938. *The petrology of the sedimentary rocks.* London: G. Allen & Unwin, 383 pp.

Hoskin, Charles M. 1963. Recent carbonate sedimentation on Alacran Reef, Yucatan, Mexico. *Natl. Acad. Sci.-Natl. Res. Council Publ. 1089*, 160 pp.

———. 1968. Magnesium and strontium in mud fraction of recent carbonate sediment, Alacran Reef, Mexico. *Am. Assoc. Petroleum Geologists Bull. 52*, 2170–77.

Howell, J. V. 1922. Notes on the pre-Permian Paleozoics of the Wichita Mountain area. *Am. Assoc. Petroleum Geologists Bull. 6*, 413–25.

Hsü, K. J., & Siegenthaler, C. 1969. Preliminary experiments on hydrodynamic movement induced by evaporation—their bearing on the dolomite problem. *Sedimentology 12*, 11–26.

Hunt, T. Sterry. 1859. On some reactions of the salts of lime and magnesium, and the formation of gypsums and magnesium rocks. *Am. Jour. Sci. Ser. 2, 28*, 377–81.

Illing, Leslie V. 1954. Bahaman calcareous sands. *Am. Assoc. Petroleum Geologists Bull. 38*, 1–95.

Illing, L. V., Wells, A. J., & Taylor, J. C. M. 1965. Penecontemporary dolomite in the Persian Gulf, *in* Pray, Lloyd C., Murray, Raymond C., (eds.), *Dolomitization and limestone diagenesis*, Soc. Econ. Paleontologists and Mineralogists Spec. Publ. *13*, 89–111.

Illing, L. V., Wood, G. V., & Fuller, J. G. C. M. 1967. Reservoir rocks and stratigraphic traps in non-reef carbonates. *Proc. 7th World Petrol. Congress, Mexico*, pp. 487–99.

Imbrie, John, & Buchanan, Hugh. 1965. Sedimentary structures in modern carbonate sands of the Bahamas, *in* Middleton, Gerald V. (ed.), *Primary sedimentary structures and their hydrodynamic interpretation*, Soc. Econ. Paleontologists and Mineralogists Spec. Pub. 12, 149–72.

Irion, Georg, & Müller, German. 1968. Mineralogy, petrology, and chemical composition of some calcareous tufa from the Schwabische Alb, Germany, *in* Müller, G., & Friedman, G. M. (eds.), *Recent developments in carbonate sedimentology in Central Europe.* New York: Springer-Verlag, pp. 157-71.

Jindřich, Vladimir. 1969. Recent carbonate sedimentation by tidal channels in the lower Florida Keys. *Jour. Sedimentary Petrology 39*, 531-53.

Johnson, Jesse Harlan. 1951. An introduction to the study of organic limestones. *Colorado School Mines Quart. 46*, no. 2, 117 pp.

————. 1961. *Limestone-building algae and algal limestones.* Colorado School of Mines, Golden, 297 pp.

Johnston, John, & Williamson, E. D. 1916. The role of inorganic agencies in the deposition of calcium carbonate. *Jour. Geology 24*, 729-50.

Kalkowsky, E. 1908. Oolith and stromatolith in Norddeutschen Buntsandstein. *Zeitschr. Deutsch. Geol. Gesell. 60*, 68-125.

Kaye, C. A. 1959. Shoreline features and Quaternary shoreline changes, Puerto Rico. *U.S. Geol. Survey Prof. Paper 317B*, 140 pp.

Kendall, Christopher G. St. C. 1969. Environmental reinterpretation of the Permian evaporite/carbonate shelf sediments of the Guadalupe Mountains. *Geol. Soc. America Bull. 80*, 2503-36.

King, Ralph H. 1947. Sedimentation in Permian Castile Sea. *Am. Assoc. Petroleum Geologists Bull. 31*, 470-77.

Klähn, H. 1928. Süsswasserkalkmagnesia Gesteine und Kalkmagnesia-süsswasser. *Chemie der Erde 3*, 453-587.

Klein, George de V. 1965. Dynamic significance of primary structures in the Middle Jurassic great oolite series, Southern England, *in* Middleton, Gerald M. (ed.), *Primary sedimentary structures and their hydrodynamic interpretation*, Soc. Econ. Paleontologists and Mineralogists Spec. Publ. 12, 173-91.

Krynine, Paul D. 1957. Dolomites (abs.). *Geol. Soc. America Bull. 68*, 1757.

Ladd, Harry S., & Schlanger, Seymour O. 1960. Drilling operation on Eniwetok Atoll, *U.S. Geol. Survey Prof. Paper 260-Y*, 863-903.

Land, L. S. 1967. Diagenesis of skeletal carbonates. *Jour. Sedimentary Petrology 37*, 914-30.

————. 1970. Phreatic versus vadose meteoric diagenesis of limestones: evidence from a fossil water table. *Sedimentology 14*, 175-85.

Land, L. S., & Epstein, S. 1970. Late Pleistocene diagenesis and dolomitization, North Jamaica. *Sedimentology 14*, 187-200.

Land, L. S., & Goreau, Thomas F. 1970. Lithification of Jamaican reefs. *Jour. Sedimentary Petrology 40*, 457-62.

Land, L. S., MacKenzie, Fred T., & Gould, Stephen J. 1967. Pleistocene history of Bermuda. *Geol. Soc. America Bull. 78*, 993-1006.

Leitmeier, Hans. 1910. Zur Kenntnis der Carbonate, Die Dimorphie des Kohlensauren Kalkes, I. Teil, *Neues Jahrb. Mineral., Heft 1*, 49-74.

————. 1915. Zur Kenntnis der Carbonate, II. Teil. *Neues Jahrb. Mineral., Beilageband. 40*, 655-700.

Lipman, C. B. 1929. Further studies on marine bacteria with special reference to the Drew hypothesis on calcium carbonate precipitation in the sea. *Carnegie Inst. Washington Publ. 391, Papers Tortugas Lab., 26*, 231-48.

Lippmann, Friedrich. 1960. Versuche zur Aufklärung der Bildungsbedingungen von Calcit and Aragonit. *Fortschr. Mineralogie 38,* 156–61.

Logan, Brian W., Harding, James L., Ahr, Wayne, M., Williams, Joseph D., & Snead, Robert G. 1969. Carbonate sediments and reefs, Yucatan Shelf, Mexico. *Am. Assoc. Petroleum Geologists Mem. 11,* 1–198.

Lowenstam, H. A. 1955. Aragonite needles secreted by algae and some sedimentary implications. *Jour. Sedimentary Petrology 25,* 270–72.

Lowenstam, H. A., & Epstein, S. 1957. On the origin of sedimentary aragonite needles of the Great Bahama Bank. *Jour. Geology 85,* 364–75.

Maiklem, W. R. 1970. Carbonate sediments in the Capricorn Reef Complex, Great Barrier Reef, Australia. *Jour. Sedimentary Petrology 40,* 55–80.

Majewskie, O. P. 1969. *Recognition of invertebrate fossil fragments in rocks and thin sections,* International Sedimentary Petrographical Series. Leiden: E. J. Brill, 101 pp.

Manheim, Frank T. 1967. Evidence for submarine discharge of water on the Atlantic continental slope of the southern United States. *New York Acad. Sci. Trans. 29,* 839–53.

Marschner, Hannelore. 1968. Ca-Mg distribution in carbonates from the Lower Keuper in northwest Germany, in Müller, G., & Freidman, Gerald M. (eds.), *Recent developments in carbonate sedimentology in Central Europe.* New York: Springer–Verlag, pp. 128–35.

Matthews, R. K. 1966. Genesis of Recent lime mud in southern British Honduras. *Jour. Sedimentary Petrology 36,* 428–54.

————. 1967. Diagenetic fabrics in biosparites from the Pleistocene of Barbados, West Indies. *Jour. Sedimentary Petrology 37,* 1147–53.

————. 1968. Carbonate diagenesis: equilibration of sedimentary mineralogy to the subaerial environment: Coral Cap of Barbados, West Indies. *Jour. Sedimentary Petrology 38,* 1110–19.

Mawson, Douglas. 1929. South Australian algal limestones in the process of formation. *Quart. Jour. Geol. Soc. London 85,* 613–21.

Maxwell, W. G. H., Day, R. W., & Fleming, P. J. G. 1961. Carbonate sedimentation on the Heron Island Reef, Great Barrier Reef. *Jour. Sedimentary Petrology 31,* 215–30.

Maxwell, W. G. H., Jell, J. S., & McKellar, R. G. 1964. Differentiation of carbonate sediments in the Heron Island Reef. *Jour. Sedimentary Petrology 34,* 294–308.

McKee, Edwin D., & Resser, C. E. 1945. *Cambrian history of the Grand Canyon Region.* Carnegie Inst. of Washington, Publ. 563, 232 pp.

Michard, Andre. 1969. Les Dolomies, une Revue. *Serv. Carte Geol. Alsace-Lorraine Bull. 22,* 1–92.

Milliman, John E. 1966. Submarine lithification of carbonate sediments. *Science 153,* 994–97.

————. 1967. Carbonate sedimentation on Hogsty Reef, a Bahamian Atoll. *Jour. Sedimentary Petrology 37,* 658–76.

Mišik, Milan. 1968. Some aspects of diagenetic recrystallization in limestones. *XXIII Int. Geol. Congress, Prague, Proc. Sect. 8,* pp. 129–36.

Moore, Hilary B. 1939. Faecal pellets in relation to marine deposits, *in* Trask, Parker D. (ed.), *Recent marine sediments.* Am. Assoc. Petroleum Geologists, Tulsa, pp. 516-24.

Moresby, Robert. 1835. Extracts from Commander Moresby's report on the northern atolls of the Maldives. *Jour. Royal Geogr. Soc. 5,* 398-403.

Murray, Raymond C. 1969. Hydrology of South Bonaire, Netherlands Antilles—a rock selective dolomitization model. *Jour. Sedimentary Petrology 39,* 1007-13.

Nason, F. L. 1901. The geological relations and the age of the St. Joseph and Potosi limestone of St. Francois County, Missouri. *Am. Jour. Sci. (4) 12,* 358-61.

Neumann, A. Conrad. 1965. Processes of Recent carbonate sedimentation in Harrington Sound, Bermuda. *Bull. Marine Science Gulf and Caribbean 15,* 987-1035.

Newell, Norman D., Purdy, Edward G., and Imbrie, John. 1960. Bahaman oolitic sand. *Jour. Geology 68,* 481-97.

Nickl, H. J., & Henisch, H. K. 1969. Growth of calcite crystals in gels. *Jour. Electrochemical Soc. 116,* 1258-60.

Perkins, R. D. 1968. Primary rhombic calcite in sedimentary carbonates. *Jour. Sedimentary Petrology 38,* 1371-73.

Pettijohn, F. J. 1949. *Sedimentary rocks.* New York: Harper and Brothers, 526 pp.

———. 1957. *Sedimentary rocks,* 2nd edition. New York: Harper and Brothers, 718 pp.

Pia, Julius. 1933. Die Rezenten Kalksteine. *Zeitschr. fur Kristallographie, Min. Pet.* (Leipzig) Abt. B, Erganzungsband, 420 pp.

Pilkey, Orrin H., Morton, Robert W., & Luternauer, John. 1967. The carbonate fraction of beach and dune sands. *Sedimentology 8,* 311-27.

Price, P. B., Vermilyea, D. A., & Webb, M. B. 1958. On the growth and properties of electrolytic whiskers. *Acta. Met. 6,* 524-31.

Purdy, Edward G. 1963. Recent calcium carbonate facies of the Great Bahama Bank: 2. Sedimentary Facies. *Jour. Geology 71,* 472-97.

———. 1968. Carbonate diagenesis: an environmental survey. *Geologica Romana 7,* 183-228.

Revelle, Roger, and Fairbridge, Rhodes. 1957. Carbonates and carbon dioxide. *Geol. Soc. America Mem. 67,* pp. 239-96.

Rose, Peter R. 1970. Stratigraphic interpretation of submarine vs. subaerial discontinuity surfaces: an example from the Cretaceous of Texas. *Geol. Soc. America Bull. 81,* 2787-97.

Russell, R. J. 1962. Origin of beach rock. *Zeitschr. Geomorphologie 6,* 1-16.

Sander, Bruno K. 1936. Beitrage zur Kenntnis der Ablagerungsgefüge (Rhythmische Kalk und Dolomite aus der Trias). *Mineralogische Petrograph. Mitt., 48,* 27-139, 141-209. (English translation by E. B. Knopf, 1951, Amer. Assoc. Petroleum Geologists, Tulsa.)

Schlanger, Seymour O. (with sections by D. L. Graf, J. R. Goldsmith, G. A. MacDonald, W. M. Sackett, & H. A. Potratz). 1963. Subsurface geology of Eniwetok Atoll. *U.S. Geol. Survey Prof. Paper, 260-BB,* pp. 991-1066.

Schmidt, Volkmar. 1965. Facies, diagenesis and related reservoir properties in the Gigas Beds (Upper Jurassic), Northwestern Germany, *in* Pray, Lloyd C., & Murray, Raymond C. (ed.), *Dolomitization and limestone diagenesis,* Soc. Econ. Paleontologists and Mineralogists Spec. Publ. 13, pp. 124–68.

Schroeder, J. H. 1969. Experimental dissolution of Ca, Mg, and Sr from recent biogenic carbonates, a model of diagenesis. *Jour. Sedimentary Petrology 39,* 1057–73.

Shearman, D. J., Khouri, J., & Taha, S. 1961. On the replacement of dolomite by calcite in some Mesozoic limestones from the French Jura. *Geologists' Assoc. Proc. 72,* 1–12.

Sherman, G. D., & Thiel, G. A. 1939. Dolomitization in glaciolacustrine silts of Lake Agassiz. *Geol. Soc. America Bull. 50,* 1535–52.

Sherman, G. D., Schultz, Florence, & Alway, F. J. 1962. Dolomitization in soils of the Red River valley, Minnesota. *Soil Sci. 94,* 309–15.

Shinn, E. A. 1964. Recent dolomite, Sugarloaf Key, Florida, *in* Ginsburg, R. N. (ed.), *South Florida Sediments,* Geol. Soc. America Annual Convention, Miami Beach, Florida, Guidebook Field Trip #1, pp. 26–33.

———. 1969. Submarine lithification of Holocene carbonate sediments in the Persian Gulf. *Sedimentology 12,* 109–44.

Shinn, E. A., Ginsburg, R. N., & Lloyd, R. M. 1965. Recent supratidal dolomite from Andros Island, Bahamas, *in* Pray, Lloyd C., & Murray, Raymond C. (eds.), *Dolomitization and limestone diagenesis.* Soc. Econ. Paleontologists and Mineralogists Spec. Publ. 13, pp. 112–23.

Skeats, Ernest W. 1903. The chemical composition of limestones from upraised coral islands, with notes on their microscopical structures. *Harvard Coll. Mus. Comp. Zoology 42,* 53–126.

———. 1905. On the chemical and mineralogical evidence as to the origin of the dolomites of Southern Tyrol. *Geol. Soc. Quart. Jour. London 61,* 97–141.

———. 1918. The formation of dolomite and its bearing on the coral reef problem. *Am. Jour. Sci. 45,* 188–89.

Sorby, H. C. 1851. On the microscopical structure of the calcareous grit of the Yorkshire Coast. *Geol. Soc. London Quart. Jour. 7,* 1–6.

———. 1853. On the microscopical structure of some British Tertiary and post-Tertiary fresh-water marls and limestones. *Geol. Soc. London Quart. Jour. 9,* 344–46.

———. 1855. On some of the mechanical structures of limestones. *Brit. Assoc. Advan. Sci. Rept. 1855 (2),* 97.

———. 1861. On the organic origin of the so-called "crystalloids" of the chalk. *Annals Magazine Natural History (London) 8, 3rd ser., no. 45,* pp. 193–200.

———. 1862. On the cause of difference in the state of preservation of different kinds of fossil shells. *Brit. Assoc. Advan. Sci. Rep. 32, pt. 2,* pp. 95–96.

———. 1879. Anniversary Address of the President: Structure and origin of limestones. *Geol. Soc. London Proc. 35,* 56–95.

———. 1904. Notes on the coral rock of Funafuti. *Royal Soc. London, Atoll of Funafuti*, 390–91.

Spörli, Bernhard. 1968. Syntectonic dolomitization in the Helvetic Nappes of Central Switzerland. *Geol. Soc. America Bull. 79*, 1839–46.

Steidtmann, E. 1911. Evolution of limestone and dolomite. *Jour. Geology 19*, 323–45, 393–428.

———. 1917. Origin of dolomite as disclosed by stains and other methods. *Geol. Soc. America Bull. 28*, 431–50.

Stockman, K. W., Ginsburg, R. N., & Shinn, E. A. 1967. The production of lime mud by algae in South Florida. *Jour. Sedimentary Petrology 37*, 633–48.

Stoddart, D. R. 1964. Carbonate sediments of Half Moon Cay, British Honduras. *Atoll Research Bull. 104*, 16 pp.

Stoddart, D. R., & Cann, J. R. 1965. Nature and origin of beach rock. *Jour. Sedimentary Petrology 35*, 243–47.

Stose, G. W. 1910. Mercersburg–Chambersburg folio. *U.S. Geol. Survey Folio 170*, 19 pp.

Stout, Wilber. 1944. Dolomites. *Ohio Jour. Science 44*, 219–35.

Swinchatt, Jonathan P. 1965. Significance of constituent composition, texture, and skeletal breakdown in some Recent carbonate sediments. *Jour. Sedimentary Petrology 35*, 71–90.

Taft, William H. 1963. Cation influence on diagenesis of carbonate sediments (abs.). *Geol. Soc. America Spec. Paper 76*, 172.

Taylor, J. C. M., & Illing, L. V. 1969. Holocene intertidal calcium carbonate cementation, Qatar, Persian Gulf. *Sedimentology 12*, 69–107.

Thomas, Carroll. 1965. Origin of pisolites. *Am. Assoc. Petroleum Geologists Bull. 49*, 45.

———. 1968. Vadose pisolites in the Guadalupe and Apache Mountains, West Texas, *in* Permian Basin Section, *Soc. Econ. Paleontologists and Mineralogists, Publ. 68-11*, pp. 32–35.

Van Tuyl, F. M. 1916. The origin of dolomite. *Iowa Geol. Survey 25*, 251–422.

Vater, Heinrich. 1899. Ueber den Einfluss der Lösungsgenossen auf die Krystallisation des calciumcarbonates. VII. Der Einfluss des Calciumsulfates, Kaliumsulfates und Natriumsulfates. *Zeitschr. Kristallographie, Kristallgeometrie, Kristallphysik, Kristallchemie 30*, 485–508.

Vaughan, T. Wayland. 1914. Preliminary remarks on the geology of the Bahamas, with special reference to the origin of the Bahama and Floridian oolites. *Carnegie Inst. Washington, Publ. 182*, pp. 47–54.

———. 1917. Chemical and organic deposits of the sea. *Geol. Soc. America Bull. 28*, 933–44.

Von der Borch, Christopher. 1965. The distribution and preliminary geochemistry of modern carbonate sediments of the Coorong area, South Australia. *Geochem. et Cosmoch. Acta 29*, 781–99.

Wade, Arthur. 1911. Some observations on the eastern desert of Egypt; with considerations bearing upon the origin of the British Trias. *Geol. Soc. London Quart. Jour. 67*, 238–61.

Walcott, C. D. 1894. Paleozoic intraformational conglomerates. *Geol. Soc. America Bull.* 5, 191–98.

Wells, A. J., & Illing, L. V. 1964. Present day precipitation of calcium carbonate in the Persian Gulf, in *Deltaic and shallow marine deposits*, 6th Int. Sedimentological Congress, 1963. Amsterdam: Elsevier. *Developments in Sedimentology 1*, 429–35.

Wethered, E. 1890. On the occurrences of the genus Girvanella in oolitic rocks and remarks on oolitic structure. *Geol. Soc. London Quart. Jour.* 46, 270–83.

Williams, Roy E. 1970. Groundwater flow systems and accumulation of evaporite minerals. *Am. Assoc. Petroleum Geologists 59*, 1290–95.

Wilson, R. C. L. 1967. Particle nomenclature in carbonate sediments. *Neues Jahrb. Geol. Paläont., Monatsh. 8*, 498–510.

Wolf, Karl H. 1965a. Gradational sedimentary products of calcareous algae. *Sedimentology 5*, 1–37.

———. 1965b. Littoral environment indicated by open-space structures in algal limestones. *Paleogeogr., Pal. and Pal. 1*, 183–223.

Wolf, Karl H., & Conolly, John. 1965. Petrogenesis and paleoenvironment of limestone lenses in upper Devonian reef rock of New South Wales. *Paleogeogr., Pal. and Pal. 1*, 69–111.

Wood, A. 1941. "Algal dust" and the finer-grained varieties of carboniferous limestone. *Geol. Mag. 78*, 192–200.

Chapter Six

Biostratinomy: The Sedimentology Of Biologically Standardized Particles

Adolf Seilacher

INTRODUCTION

Because the ultimate goal of sedimentology is reconstructing ancient environmental conditions, the traditional link with paleontology is in the field of environmental paleontology, or *paleoecology*. Much work has been done in recent years following the basic idea that the same environmental factors may control both the sedimentary processes and the behavior, shape, and distribution of contemporary organisms. The interrelationship, however, continues after the death of these organisms, though under the complete dominance of inorganic processes that tend to bring the biogenic skeletal particles into physical and chemical equilibrium with the surrounding sediment (Böger 1970).

The post-mortem history of fossils is the subject of *taphonomy* (Efremov 1940), which can be further subdivided into *biostratinomy* (Weigelt 1919), dealing with the sedimentational history of organic remains, and *fossilization* (or "Fossil-Diagenese," A. H. Müller 1951), comprising their alterations after the final burial. With regard to the hard parts, biostratinomic processes are mainly physical, while fossilization processes are largely chemical in nature.

Biostratinomy was developed mainly by paleontologists, who have a vital interest in removing the "taphonomic overprint" (Lawrence 1968) before they start the paleobiological interpretation of the fossil record. But the field could with equal right be claimed by the sedimentologists, who should be interested to study the behavior of standardized particles that are morphologically and structurally much more differentiated than sand grains and still occur often in large enough numbers to warrant statistical and experimental evaluation. Considering the potentials in comparison to the little work done in this field, the present review is hoped to be prospective as much as retrospective in character.

I. THE FOUNDERS OF BIOSTRATINOMY

A short look at the history of the field is necessary to understand its present state. Without exception the founders came from the biomorphological rather than the physicochemical wing of geology, and all got their initial impetus from observations in present-day environments. It is also interesting to note that for some reason their new way of looking at fossils did not spread easily. As a result, biostratinomy remained a somewhat endemic science centered in two schools—both of them in Germany, and both mainly paleontological.

The Halle school goes back to Johannes Walther (1860–1937; details in Weigelt 1927). Walther wrote a zoological dissertation and worked on faunal distributions in the Gulf of Naples, Italy, where his teacher Ernst Haeckel had earlier discovered the beauty of radiolarian skeletons. But the fauna-sediment relationships and subsequent travels through the deserts of North Africa and to Red Sea coral reefs turned his interest so much toward geology that he became a devoted geology professor, known through the rule of facies succession and through important books on geology as a historical science (1890-93), on desert formation (1900), and on general paleontology (1927). Drawing on a broad spectrum of his own observations, he pioneered the use of present-day environments as the key to the past. It is a strange coincidence that Walther also spent some time as a guest professor at The Johns Hopkins University, though many years before Francis Pettijohn began working along similar lines there.

Walther's pupil Johannes Weigelt (1890–1948), see Voigt (1962), in contrast, started from the geological side. He first conceived the potential of present-day environments for interpreting ancient deposits during a stay at the Wadden Sea (Weigelt, 1919). Later when he worked with a seismic crew in Texas he was so impressed by the mass mortality caused by a winter storm that he summarized his observations in a book on recent vertebrate corpses and their paleobiological significance (Weigelt, 1927). This book, though largely concerned with articulated vertebrate remains, marks the true beginning of "biostratonomy" a term that for etymological reasons has later been changed to "biostratinomy" (Wolf 1954). After he succeeded Walther, Weigelt successfully applied biostratinomy to his excavations in the Eocene brown coal of the Geiseltal, to his work on the vertebrate fossils of the Mansfeld copper shales (Permian), and in the prospection of the Salzgitter iron ore deposits, all in central Germany. During the third generation of the Halle school, biostratinomic principles, now applied to invertebrate as much as vertebrate remains, were summarized in A. H. Müller's booklet (1951) and expanded in his textbook of paleontology (Müller 1963).

The Frankfurt tradition in biostratinomy started with Rudolph Richter (1881–1957), who founded the Wilhelmshaven Marine Station of the Senckenberg Institution as a center of what he called "Aktuopaläontologie."

This program marked a change in emphasis from vertebrate to invertebrate remains and from continental to marine environments. The tremendous amount of observational data collected by Richter and his collaborators (Schwarz 1942; Hecht 1933) called for comparison with fossil examples, but also tended to over-emphasize the Wadden Sea model in paleoenvironmental interpretations. Richter's work was very successfully carried on by W. Schäfer, whose book (1962) summarizes and beautifully illustrates vertebrate and invertebrate biostratinomy in the biological and physical framework of a modern depositional environment (Fig. 1).

II. SEDIMENTARY TRANSPORT

Because the original attitudes, shapes, and relative frequencies of many fossils are known, their present orientation, fragmentation, and sorting may serve as a clue to the kind, direction, and intensity of their post-mortem sedimentary transport.

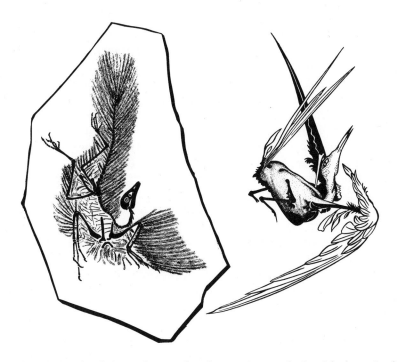

Fig. 1. Much of the earlier work in biostratinomy dealt with the attitude of articulated vertebrate skeletons. *Archaeopteryx* can be compared to mummified Recent birds, in which the sharp dorsal bent of the vertebral column is caused by post-mortem contraction of dorsal muscles and tendons (after Colbert 1955, Fig. 57, and Schäfer 1955, Fig. 2).

a. Orientation

Few fossils are perfect spheres; most show various degrees of asymmetry that enable us to describe, measure, and compare their orientations much more precisely than the orientation of sand grains. Only in rare cases will this orientation be random. In fact, alignment is so generally accepted that *chaotic fossil orientation* has only recently been recognized as a possible indication of exceptional processes, such as slumping or complete bioturbation (Toots 1965). In most cases fossils are aligned, at least with their long axes parallel to the bedding plane. Therefore it is usually sufficient to describe their orientation within this plane (Seilacher 1959).

Perhaps the most familiar orientation phenomenon is the preferred *convex-up* position of bowl-shaped bodies (ostracode, pelecypod, and brachiopod valves, arthropod carapaces, etc.) exposed to tractional currents ("Einkippungs-Regel," Richter 1942). If the opposite, *convex-down* position prevails, it indicates either lack of currents or vertical sedimentation in sediment traps (Fig. 10) or as flotsam (Richter 1922, p. 106).

Preferred *azimuth orientations* ("Einsteuerung," Richter 1942), most obvious in elongate fossils (graptolites, tentaculites, orthocone cephalopods, high-spired gastropods, etc.) or in flexible appendages, may also serve to determine the direction of paleocurrents. Measurements, taken if necessary from a larger surface or from different bedding planes, can be combined in a rose diagram, the shape of which distinguishes between biological or mechanical *current* orientation and *wave* orientation (Fig. 2; Nagle 1967). This distinction is useful not only for correct paleocurrent determination but also for a gross bathymetric interpretation, as Nagle (1967) has demonstrated by field measurements in the Devonian of the Appalachians (Fig. 3).

The two specimens of Figure 4 are interesting in spite of the fact that their original orientation in the field is unknown and the current arrows can therefore not be drawn in a map. But the demonstration of currents as such was important in both cases, as the fine black sediments would easily have been interpreted as typical quiet-water deposits (V. Koenigswald 1930; Richter 1931; Seilacher 1960).

Bundenbach starfishes (Fig. 4) also help to understand our rose diagrams more *dynamically*: As in the orthocone slab (Fig. 4), there are always a few "black sheep" pointing in the opposite direction, although this does not necessarily correspond to a preferred position in terms of stability. Formerly, these counterorientations were ascribed to flood and ebb currents in a tidal flat environment (v. Koenigswald 1930, p. 357), but they can also be explained as stages in a unidirectional rolling transport, during which the arms are subsequently turned over the body or dragged underneath. The probability of preserving each position would depend on its average *residence time* within this movement.

CURRENT ORIENTATION

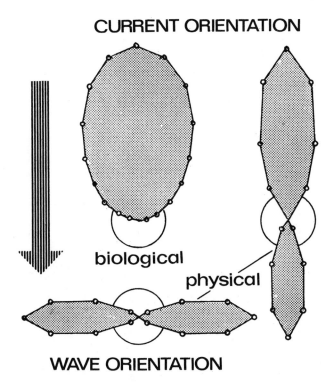

biological

physical

WAVE ORIENTATION

Fig. 2. Rose diagrams help to recognize and distinguish various sorts of alignment of fossils on bedding planes. *Biological current orientation*, common in current feeders, is expressed by a broad peak, but lacks counterorientations. *Physical current orientation* of elongate particles produces a sharp peak and a smaller counterpeak. Transverse *wave orientation* is characterized by two equal peaks, sometimes with an additional smaller peak in the direction of wave progression (after Seilacher 1960, Fig. 6).

Groups of oriented fossils can be considered frozen stages of dynamic transportation. To understand how groups of only a few fossils can have differing orientations we should use serial photography of natural environments or experiments in flumes. The experimental approach might be particularly fruitful because it would permit us to vary the forms and the abundance of tests as well as the sediment and current velocity. Systematic *experimental* studies might eventually enable us to reconstruct not only the direction but also the velocity of paleocurrents.

As a bedding plane gets crowded with shells, mutual *inhibition* gradually overshadows the current-induced azimuth orientations. Crowding, combined with continued agitation, will, on the other hand, facilitate the formation of

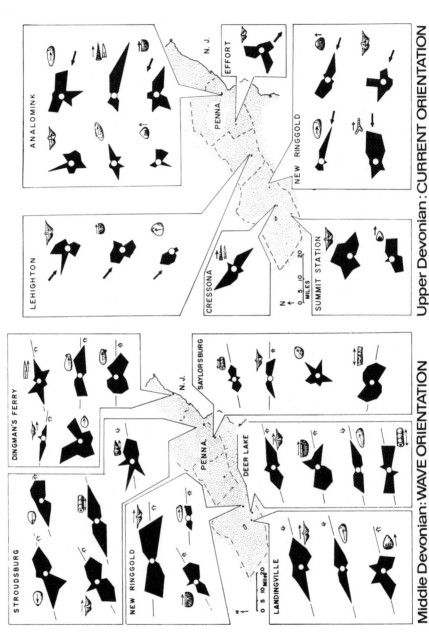

Middle Devonian: WAVE ORIENTATION   Upper Devonian: CURRENT ORIENTATION

Fig. 3. Transition from dominant wave to current orientation marks the deepening below wave base of the

edgewise coquinas (Fig. 5). Shells or other flat objects can obtain very long residence times by being wedged between stationary pebbles or vertically implanted shells. "Stacks" of shells in which many valves snugly fit into each other are another expression of the same phenomenon. The only mechanism needed is a turbulent environment providing a very high number of trials, such as a pebbly mud beach on which edgewise coquinas form today. It is probable that fossil examples (Fig. 5) also formed on very shallow bottoms dominated by wave action.

b. Fragmentation

The standard shapes and structures of biogenic particles as compared with inorganic particles make not only their transport but also their physical destruction more predictable. It has long been known that predators, such as boring gastropods or shell-breaking crustaceans (Tauber 1947), may produce distinctive fragments, but different sedimentological processes should also be

Fig. 4. Current orientation. Preferred orientations (counterpositions drawn in black) can be interpreted as phases of current transport. Their representation corresponds to their relative residence times (after Seilacher 1960).

reflected in particular kinds of breakage and abrasion (Fig. 6). Fossil examples of rolled fragments are shown in Figures 7 and 8, and several other occurrences could be quoted. The fact that most paleontologists are much more interested in complete fossils than in fragments has caused our general lack of pertinent information. More systematic observations in ancient and recent sediments (Driscoll 1967; Hollmann 1968; Rolin 1971) and in experimental tanks are needed before we can use not only the quantitative (Chave 1964) but also the morphological aspect of skeletal durability for environmental interpretations.

c. Sorting

Sorting is well known to distort the original size distribution in shell assemblages (Fagerstrom 1964; Craig 1967), but little has been done to evaluate this distortion for sedimentological purposes. Crinoids (Fig. 15) are best suited for studying sorting because they produce a regular number of diverse skeletal elements.

Beds of disarticulated shells with an unequal proportion of both valves is the rule rather than the exception. This inequality is easy to explain if one valve is more fragile than the other. In other examples, differing morphology of the two valves can lead to variation in flow resistance, rollability, or anchoring, eventually effecting separation of the two valves during bottom transport. Brachiopods with teeth only in the pedicle valve and *Mya* having a protruding chondrophor only in the left valve are familiar examples (Richter 1922, p. 130).

In modern beaches, however, even perfectly equal valves, for instance of *Arca, Pitar,* or *Donax,* are often found in significantly different numbers (Martin-Kaye 1951; Craig 1967). Following earlier suggestions (Gressly 1861), the Dutch zoologist J. Lever has studied this phenomenon extensively in the field with real and artificial shells (1958, 1961, 1964). He was able to show that the right-left symmetry of the valves in combination with a right-left symmetry in the hydrodynamic system of the surf can actually cause the observed separation (Fig. 9).

III. INTERNAL SEDIMENTATION

Many fossils have acted not only as sedimentary particles but also as sediment recipients before they became buried. Biogenic cavities can certainly not compare in size with what we usually call sedimentary basins, but they have the considerable advantage of defined configuration and multiple occurrence.

Depending on their shapes, apertures, and orientations, as well as the hydrodynamic situation, biogenic sediment traps may provide important information for the sedimentologist. For example, they often retain material that did not come to rest on the level bottom, because it either rolled along

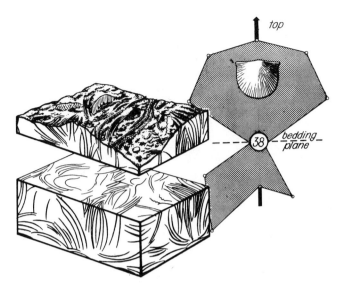

Fig. 5. In edgewise coquinas, flat valves are wedged between obstacles or other shells by highly agitated litoral waters. The block (Ordovician, Cincinnati) shows packages opening like books on a basal flat-shell layer, but lacking over-all alignment in the horizontal section. In the rose diagram, taken from vertical sections through a similar brachiopod coquina in the Silurian of Norway, a spade-like orientation with the umbo pointing upward prevails (from Seilacher & Meischner 1964, Fig. 6).

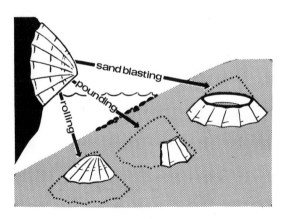

Fig. 6. Fragmentation. Abrasion and breakage may produce diagnostic fragments in different environments. If the shell remains fairly stable (muddy bottoms or dune sand), abrasion attacks only the top side (Pratje 1929, "Fazettierung"). Breakage on a rocky shore will largely follow radial and concentric shell structures, while rolling and shifting will affect mainly the free edge.

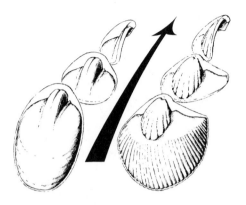

Fig. 7. Roll fragments. Most brachiopod valves in the Oriskany Sandstone of Berkeley Springs (W. Va.) have lost their thinner marginal parts through roll abrasion. The remaining umbonal portion is internally thickened and bears an oversized muscle scar indicating the size of the unabraded shell (after Seilacher 1968, Fig. 3).

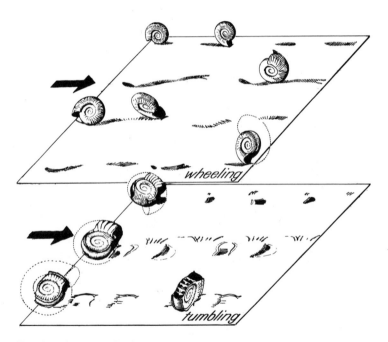

Fig. 8. Abrasion of rolling ammonite shells. Many of the roll marks in the Solnhofen lithographic limestones (U. Jurassic) can be referred to ammonite shells that were distinctively worn by the rolling process. At a certain stage of destruction, transition from wheeling to lateral tumbling has also changed the type of shell wearing (from Seilacher 1963, Figs. 2-3).

or remained in suspension (Fig. 10). If the internal sedimentation was incomplete, the remaining voids can be used as geopetal structures indicating tops and bottoms in folded beds, reworked blocks, or float specimens. Broadhurst and Simpson (1967) have shown that they may also help to distinguish paleoslope from secondary tilting (Fig. 11).

It seems, however, that the sediment fill may also give a clue to the particular hydrodynamic situation in which it was deposited. The chambered parts of ammonite shells, for instance, remained consistently unfilled in certain facies, while sedimentary steinkerns dominate in others. The fillings may also have their own *sedimentational history*. Often they start with coarser material that fines upwards—a tendency that would probably become clearer if studied under higher magnifications. Not only the grain size but also the depositional structures may be graded in the sediment fill. This is particularly clear in cavities, such as certain burrows (Fig. 12) or ammonite chambers, that once had constricted openings and therefore became filled by an internal draft current set up by outside turbulence. The chambers of a ceratite phragmocone, for instance, fill up with even layers to about the level of the siphuncular passages. But the sedimentation inside continues beyond that level, tending to form concave or even bubble-shaped surfaces that finally

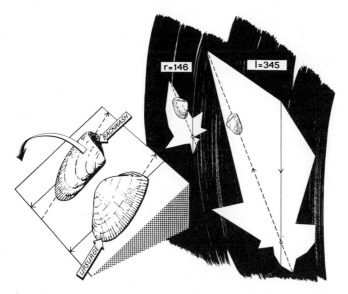

Fig. 9. In Recent beaches the swash orients right and left valves of equivalve pelecypods, such as *Donax*, to mirror-image rose diagrams, but the backwash tends to flip or rotate them both in opposite directions. The left-right sorting (unequal numbers!) becomes more efficient if the shell asymmetry is combined with a left-right asymmetry of the hydrodynamic system (after Lever 1958).

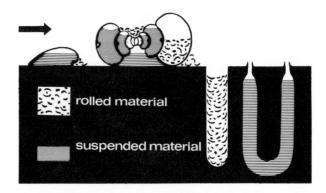

Fig. 10. Biogenic sediment traps. Depending on the hydrodynamic situation and on the width of the openings, biogenic cavities remain empty, or become gradually filled with material that is often coarser or finer than the surrounding sediment. In flat-lying ammonite shells (after Merkt 1966, Fig. 5) internal sedimentation may differ between body and gas chambers and between upper and lower umboes.

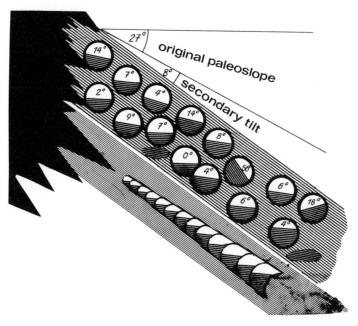

Fig. 11. Voids as paleoslope indicators. In limestone beds dipping steeply away from a Silurian reef complex the fillings of shells and sediment cavities make it possible to distinguish the paleoslope component from compactional tilting (after Broadhurst & Simpson 1967, fig. 1).

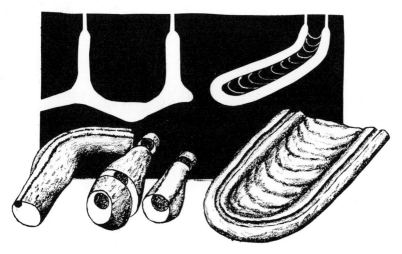

Fig. 12. Draft filling. Cavities with constricted openings, such as burrows of the *Thalassinoides* (*left*) and *Rhizocorallium* type (*right*), often become filled by an internal draft current set up by external turbulence. A tube-like fill channel, equal to the openings in diameter and running along the crest of the burrow fill, marks the end of draft action. It remains either open or becomes filled with material differing from the other sediment fill (from Seilacher 1968, Fig. 5).

---

close around a continuous fill channel (Fig. 13). This channel is particularly striking because it has the diameter of the siphuncular openings which it connects and because it remains either open or fills with different sediment, indicating an abrupt change in the dynamics of internal sedimentation (Seilacher 1967).

IV. PREFOSSILIZATION

In many neritic sediments fossils rarely remain undisturbed at the place of their first burial. Most are reworked, many after enough time in the sediment to have undergone early diagenetic alterations. They re-enter sedimentation unchanged in shape but different in durability and specific weight and accordingly different in behavior.

Most "prefossilized" remains are friable and disintegrate very rapidly, but others, particularly if they originally consisted of porous material, are much more durable than they were before. They are heavier and harder through mineralization of the pore spaces, so that reworked coprolites, bone fragments, and teeth are concentrated and sorted together with coarse sand grains to form the typical *bone beds* (Fig. 14). Their previous petrification is further proven by pre-burial angular fractures of the formerly spongy or soft bones and coprolites (Reif 1971) or by bone fragments that sedimentation has polished like gems in a tumbler.

The spongy stereome of the *echinoderms* is in a way similar to vertebrate bone, so that prefossilization would drastically change the sedimentological behavior of echinoderm particles (Fig. 15). Layers of "tumbled" crinoid columnals have in fact been found (Linck 1965, p. 141), but for some reason these seem to be much rarer than primary concentrations (Ruhrmann 1971).

Prefossilization may also increase the durability of *mollusc shells.* In this case it is the enclosed sediment fill that tends, by concretionary processes, to harden earlier than the surrounding mud. Prefossilized steinkerns often survive subsequent resedimentation so perfectly that their reworked nature can only be demonstrated by internal fill structures inconsistent with their present orientation in the rock (Fig. 16). In Muschelkalk ceratites it seems that reworked specimens are at least as common as shells in primary burial posi-

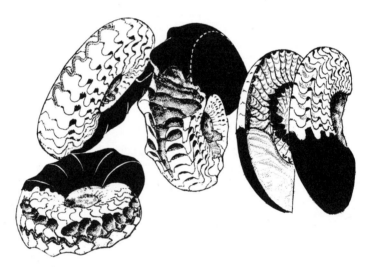

Fig. 13. Fill channels and lobe voids. Draft through siphuncular necks and punctures is in many cases responsible for the sedimentations inside cephalopod phragmocones. In shells filled in horizontal or inclined position the fill channel bends up between the septa to form a wavy line, while it is straight and touches only the upper surfaces in vertical specimens. Lobe voids (*drawn in black*) left open in the upper corners of the corrugated septal roofs allow reconstruction of the fillings' positions of the ceratite shells illustrated, which had all been reworked and redeposited in horizontal position, but with their fillings already prefossilized. The lamination seen in the median section shows that all chambers became filled synchronously by concave layers of very fine sediment (from Seilacher 1968, Figs. 1 and 3).

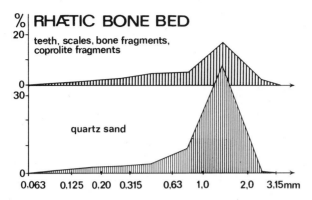

Fig. 14. Teeth, scales, and bone or coprolite fragments conform with the histogram of the quartz sand fraction despite original differences in specific weight, consistency, and size. This fact and the angular shape of ancient fracture surfaces indicates that the vertebrate remains were already prefossilized by the time of bonebed sedimentation (data by courtesy of W. Reif, Tübingen).

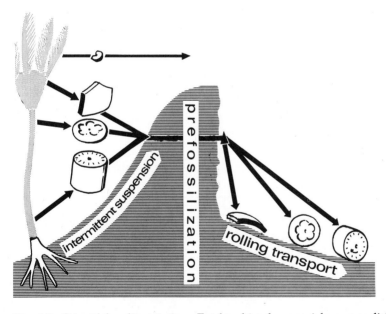

Fig. 15. Crinoidal sedimentation. Fresh echinoderm ossicles are so light that they are carried away mainly in suspension, suffering practically no abrasion and very little sorting except by size. After prefossilization has mineralized the pore spaces in the calcite skeleton, however, bottom transport may separate and abrade the ossicles according to shape and rollability (data from Ruhrmann 1971 and personal communication).

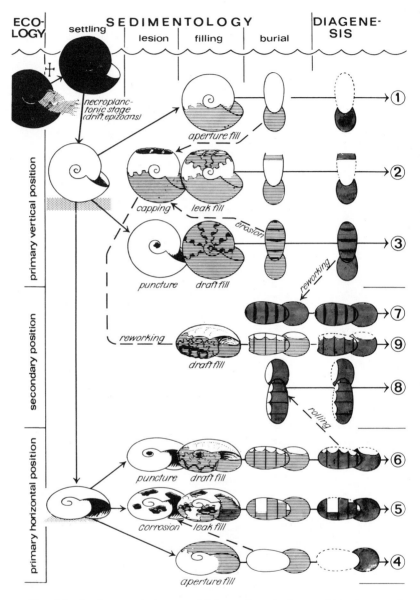

Fig. 16. A characteristic set of biostratinomical case histories can be established for every type of facies. The preservation of Muschelkalk ceratites, for instance, is mainly controlled by rapid sedimentation and frequent reworking in a shallow marine environment combined with early concretionary hardening of mud fills and solution of the aragonite shells in the calcareous mud (from Seilacher 1971).

tion. It may well be that more studies on this line will help us to better appreciate the amount and the duration of intraformational recycling, particularly in shallow marine sediments.

REFERENCES

Böger, H. 1970. Bildung und Gebrauch von Begriffen in der Paläoökologie. *Lethaia 3*, 243–69.

Broadhurst, F. M. & Simpson, I. M. 1967. Sedimentary infillings of fossils and cavities in limestone at Treak Cliff, Derbyshire. *Geol. Magazine 104*, 443–48.

Chave, K. E. 1964. Skeletal durability and preservation *in* Imbrie, J., and Newell, N. (eds.), *Approaches to paleoecology*. Wiley, New York, pp. 377–87.

Colbert, E. H. 1955. *Evolution of the vertebrates.* New York: Wiley, London: Chapman & Hall, XIII, 479 pp.

Craig, G. Y. 1967. Size-frequency distributions of living and dead populations of pelecypods from Bimini, Bahamas, B.W.I. *Jour. Geology 75*, 34–45.

Driscoll, E. G. 1967. Experimental field study of shell abrasion. *Jour. Sedimentary Petrology 37*, 1117–23.

Efremov, I. A. 1940. Taphonomy, a new branch of paleontology. *Akad. Nauk. S.S.S.R. Biul., Biol. Ser. 3*, 405–13.

Fagerstrom, J. A. 1964. Fossil communities in paleoecology: their recognition and significance. *Geol. Soc. America Bull. 75*, 1197–1216.

Gressly, A. 1861. *Erinnerungen eines Naturforschers aus Süd-Frankreich.* Zürich: Album de Combe-Varin, pp. 201–96.

Hecht, F. 1933. Der Verbleib der organischen Substanz der Tiere bei meerischer Einbettung. *Senckenbergiana 15*, 165–249.

Hollmann, R. 1968. Zur Morphologie rezenter Mollusken-Bruchschille. *Paläont. Zeitschr. 42*, 217–35.

Johnson, R. G. 1957. Experiments on the burial of shells. *Jour. Geology 65*, 527–35.

Koenigswald, R. von 1930. Die Arten der Einregelung ins Sediment bei den Seesternen und Seelilien des unterdevonischen Bundenbacher Schiefers. *Senckenbergiana 12*, 338–60.

Lawrence, D. R. 1968. Taphonomy and information losses in fossil communities. *Geol. Soc. America Bull. 79*, 1315–30.

Lever, J. 1958. Quantitative beach research. I. The "left-right-phenomenon": sorting of Lamellibranch valves on sandy beaches. *Basteria, Tijdschr. Nederland. Malacol. Verein. 22*, 21–68.

Lever, J., Kessler, A., Overbeeke, A. P. van, & Thijssen, R. 1961. Quantitative beach research. II. A second mode of sorting of Lamellibranch valves on sandy beaches. *Netherlands Jour. Sea Research 1*, 339–58.

Lever, J., Bosch, M. van den, Cook, H., Dijk, T. van, Thiadens, A. J. H., Thijssen, S. J., & Thijssen, R. 1964. Quantitative beach research. III. An experiment with artificial valves of *Donax vittatus. Netherlands Jour. Sea Research 2*, 458-92.

Linck, O. 1965. Stratigraphische, stratinomische und ökologische Betrachtungen zu *Encrinus liliiformis* Lamarck. *Jahresh. geol. Landesamt Baden-Württemberg 7*, 123-48.

Merkt, J. 1966. Über *Euagassiceras resupinatum* (SIMPSON), Ammonoidea, aus der Sauzeanumbank Nordwestdeutschlands. *Geol. Jahrbuch 84*, 23-88.

Müller, A. H. 1951. Grundlagen der Biostratonomie. *Abhandl. Deutsch. Akad. Wiss. Berlin, Kl. Math. u. allg. Naturwiss.*, Jahrg. 1950, No. 3, 147 pp.

Nagle, J. S. 1967. Wave and current orientation of shells. *Jour. Sedimentary Petrology 37*, 1124-38.

Pratje, O. 1929. Fazettieren von Molluskenschalen. *Palaeont. Zeitschr. 11*, 151-69.

Reif, W. 1971. Bonebeds an der Muschelkalk-Keuper-Grenze in Ostwürttemberg. *Neues Jahrb. Geol. Palaeont.* Abh. *139*, 369-404.

Richter, R. 1922. Flachseebeobachtungen zur Paläontologie und Geologie. III-VI. *Senckenbergiana 4*, 103-41.

———. 1931. Tierwelt und Umwelt im Hunsrückschiefer; zur Entstehung eines schwarzen Schlammsteins. *Senckenbergiana 13*, 299-342.

———. 1942. Die Einkippungsregel. *Senckenbergiana 25*, 181-206, 404.

Rolin, M. F. 1971. Etat des restes de lamellibranches dans les thanatocenoses et relations avec les conditions de formation: les cassures, *Trav. Laborat. Paleont.* Univ. Paris.

Ruhrmann, G. 1971. Riff-ferne Sedimentation unterdevonischer Krinoidenkalke im Kantabrischen Gebirge (Spanien). *Neues Jahrb. Geol. Palaeont., Monatsh.* 1971, 231-48.

Schäfer, W. 1955. Fossilisations-Bedingungen der Meeressäuger und Vögel. *Senckenbergiana Lethaea 36*, 1-25.

———. 1962. *Aktuo-Paläontologie nach Studien in der Nordsee.* Frankfurt a/Main: Kramer, VIII, 666 pp.

Schwarz, A. 1932. Der tierische Einfluss auf die Meeressedimente. *Senckenbergiana 14*, 118-72.

Seilacher, A. 1959. Fossilien als Strömungsanzeiger. *Aus der Heimat, Öhringen (Württemberg) 67*, 170-77.

———. 1960. Strömungsanzeichen im Hunsrückschiefer. *Notizbl. Hess. Landesamt Bodenforsch. Wiesbaden 88*, 88-106.

———. 1963. Umlagerung und Rolltransport von Cephalopoden-Gehäusen. *Neues Jahrb. Geol. Paläont., Monatsh.* 1963, 593-615.

———. 1967. Sedimentationsprozesse in Ammonitengehäusen. *Akad. Wiss. Literatur, Abhandl. Math.-Naturwiss. Kl. 1967*, 191-203.

———. 1968. Origin and diagenesis of the Oriskany sandstone (Lower Devonian, Appalachians) as reflected in its shell fossils, *in* Müller, G., and Friedman, G. M. (eds.), *Recent developments in carbonate sedimentology in Central Europe.* Berlin: Springer, pp. 175–85.

———. 1971. Preservational history of Ceratite shells. *Palaeontology 14,* 16–21.

Seilacher, A., & Meischner, D. 1964. Fazies-Analyse im Paläozoikum des Oslo-Gebietes. *Geol. Rundschau 54,* 596–619.

Tauber, A. F. 1947. Lebensspuren von Krebsen an fossilen Scaphopoden-schalen, *in* Papp, Zapfe, Bachmayer, & Tauber, Lebenspuren mariner Krebse. *Sitzungsber. Akad. Wiss. Wien, Math.-naturwiss. Kl. 155,* 281–317.

Toots, H. 1965. Random orientation of fossils and its significance. *Contr. to Geol., Univ. Wyoming 4,* 59–62.

Voigt, E. 1962. Johannes Weigelt als Paläontologe. *Mitteilung. Geol. Staatsinst. Hamburg 31,* 26–50.

Walther, J. 1890–93. *Einleitung in die Geologie als historische Wissenschaft.* V. 1–3. Jena: G. Fischer, 1055 pp.

———. 1912. *Gesetz der Wüstenbildung in Gegenwart und Vorzeit.* Leipzig: Quelle & Meyer, IX, 342 pp.

———. 1919. *Allgemeine Palaeontologie.* Berlin: Borntraeger, X, 548 pp.

Weigelt, J. 1919. Geologie und Nordseefauna. *Der Steinbruch 14,* 228–31, 244–46.

———. 1927. Wirbeltierleichen in Gegenwart und geologischer Vergangenheit. *Natur und Museum 57,* 97–106.

Wolff, E. 1954. Taxionomie, Stratigraphie und Stratinomie (nicht Taxonomie, Stratographie und Stratonomie) und Verkürzungen wie Palichnologie, Palökologie. *Senckenbergiana Lethaea 35,* 115–17.

Chapter Seven

Recent and Ancient Algal Stromatolites: Seventy Years of Pedagogic Cross-Pollination*

PAUL HOFFMAN

From the very beginning, the interpretation of ancient stromatolites has been dependent on Recent analogs. This was a surprise to me, for there is a tendency for those of us under thirty to believe that Recent sediments were "discovered" sometime after 1960, thus rescuing sedimentology from at best unsung decades of sieving and heavy minerals. (Having already become insulting, I had best offer the warning that when you get someone who saw his first stromatolite only five years ago to write on "Seventy years of . . . ," you can't expect sagacity. Furthermore, I expect to be forgiven for exercising the reviewer's prerogative of occasionally referring to my own inestimably worthy but sadly unpublished ideas.) From the stromatolite case history, I read the moral that when the cross-pollination of research in the ancient and the Recent is impeded the adaptive capacity of both is reduced.

CORALLINE ALGAE AS RECENT ANALOGS OF ANCIENT STROMATOLITES

By the turn of the century, there was a consensus that the various laminated calcareous growth structures, first described by James Hall (1883) and shortly to be given the name *stromatolith* by Ernst Kalkowski (1908), were biogenic. Field work in the Appalachian and Cordilleran regions showed that they are a dominant element of the lower Paleozoic and Proterozoic fossil record. But without a Recent analog geologists were powerless to decide (although there were many suggestions) what kind of organism was to be held responsible.

*This chapter is not an objective review of the stromatolite literature. The bias toward North American work is deliberate. Of the many persons with whom I have discussed stromatolite problems, I wish to acknowledge especially Robert Ginsburg, Conrad Gebelein, Brian Logan, Hans Hofmann, and Mikhail Semikhatov.

A possible Recent analog was soon found in the wake of the famous *Siboga* expedition to study the calcareous algae of Malaysia (Weber van Bosse & Foslie 1904). Eliot Blackwelder (1913), then working on the Ordovician Bighorn Dolomite of Wyoming, was one of many impressed by the resemblance of ancient stromatolites to the crustose and complexly-branching coralline algae, among the most important of Recent reef-builders. Regrettably, coralline algae were known to be abundant in polar, as well as tropical, oceans and thus they held little promise as paleo-climatic indicators. This deficiency did not detract much from their supposed geologic importance, however, and Wieland (1914), who described the proterozoic through Ordovician as the "reign of algae," was moved to write (p. 248): "Nor does it even seem too much to say that no dominant organisms of later ages whether plant or animal ever exceeded the Paleozoic seaweeds as rock-forming agents or left a bulkier record."

STROMATOLITES AS FRESH-WATER ALGAL TUFA

It was decided that algae were the culprits, but not everyone was happy in singling out the corallines. Coralline algae have a distinctive, lattice-like, skeletal microstructure, something never found in even the best preserved ancient stromatolites. As fossils with the microstructure of coralline algae became well known, it was clear that stromatolites were something different. Feedback from the ancient ruled that the first Recent analog of stromatolites was to be abandoned.

There was to be little time for mourning. In 1914 Charles Walcott provided an alternative Recent analog. Walcott compared the Proterozoic stromatolites of the Belt Supergroup in Montana with fresh-water calcareous *tufa* formed by nonskeletal blue-green algae. Tufa nodules and encrusting growth structures, well known at the time in the lakes and rivers of Pennsylvania and New York State (Roddy 1915), are formed by in situ precipitation of calcium carbonate, presumably as a result of algal photosynthesis. Like ancient stromatolites, algal tufa has concentric laminations and filament moulds, but lacks the orderly skeletal microstructure of coralline algae.

The algal tufa analogy appealed to Walcott also for another and, in retrospect, bizarre reason. He believed that the appearance of complex shelly invertebrates at the beginning of the Paleozoic was not due to their sudden evolution but was a result of the initial world-wide marine transgression of the continents. He argued that the early evolution of animals took place slowly in distant ocean basins (but not too slowly, for the entire Precambrian was then thought to represent less than half of geologic time). The fossil evidence of this evolution he believed to be lacking because the Proterozoic sedimentary rocks of the continents are entirely nonmarine. A fresh-water origin for Proterozoic stromatolites suited this theory admirably.

The fresh-water interpretation was reinforced by Wilmot Bradley's (1929) beautifully illustrated comparison of Eocene stromatolites in the lacustrine Green River Formation of Wyoming and Recent algal tufa in Green Lake, upstate New York. But the reliability of stromatolites as paleo-salinity indicators received a jolt with Armand Eardley's (1938) description of yet another lacustrine algal tufa, this one in the hypersaline Great Salt Lake of Utah. Eardley's discovery no doubt pleased European geologists, who kept finding ancient stromatolites growing on ammonites and in other equally indisputably marine associations. The Recent analogs had convinced everyone that stromatolities are built by nonskeletal blue-green algae, but the lack of Recent marine stromatolites was increasingly embarrassing.

STROMATOLITES BUILT BY SEDIMENT-BINDING ALGAL MATS

In the spring of 1930 Maurice Black, a member of Richard M. Field's International Expedition to the Bahamas, canoed the tortuous and uncharted system of tidal channels and lakes across Andros Island to the now-famous tidal flats of the Great Bahama Bank. There, he (1933) found extensive areas of Recent marine carbonate sediment with a variety of laminated algal growth structures very different from fresh-water algal tufa. They are composed of sediment particles, mainly pellets of aragonitic mud and shell fragments, bound firmly by a thin but cohesive meshwork or mat of microscopic blue-green algal filaments. In contrast to stony algal tufa, the Bahamian structures are unlithified and there is little or no in situ carbonate precipitation. Black was careful to point out that the structures are not monospecific, as in much algal tufa, but are produced by heterogeneous algal communities. Black intended to make a comparison of the Bahamian algal structures and ancient stromatolites, but this was never published. If it had been, perhaps it would have taken thirty years for the importance of his observations to become generally known.

The analogy between Recent sediment-binding algal mats and ancient stromatolites was rediscovered in south Florida by Robert Ginsburg (1955) of the enormously influential Shell Development Company laboratory in Miami. Like Black, Ginsburg challenged the traditional view that stromatolites grow by in situ precipitation. He also argued, on the basis of Recent analogs, that differences in stromatolite morphology are related to physical factors, such as wave action, sediment influx, and desiccation rather than to the biological make-up of the algal mats. Detached biscuit-like stromatolites are found in shallow turbulent water, whereas raised concave plates occur on rarely flooded supratidal mud flats. If the analogies proved correct, it was clear that stromatolites would have more potential as paleo-environmental indicators than as zone fossils for biostratigraphic correlation.

The large columnar stromatolites so typical of the Proterozoic had nowhere been found in the Recent until Brian Logan (1961) reported the still-unique occurrence in Shark Bay on the west coast of Australia. Logan was studing the foraminifera of the bay for his doctoral dissertation when he found the kilometers of stony calcareous growth structures being built by sediment-binding algal mats that occur in the intertidal zone of the hypersaline bayheads. These structures, destined to provide most of what is now known of the environment and formation of marine stromatolites, range in size from flat-laminated beds with scattered digitate clusters a few centimeters in diameter (in protected embayments), through elongate domes and narrow complexly-branching columns with several centimeters relief (in semiprotected bights), to discrete club-shaped columns and mound-like columnar aggregates with relief of a meter and more (at exposed headlands). Logan was not the first geologist to see these structures, but as he had just completed a predoctoral thesis on some Proterozoic stromatolitic cherts he immediately realized that these were by far the most convincing Recent analogs ever found.

In the aftermath of the Shark Bay discovery, the ideas earlier expressed by Black and Ginsburg became widely accepted. The sediment-binding origin and tidal-flat environment of stromatolites became almost a cliché, as is evident from the flood of papers on ancient carbonate tidal flats published during the late sixties. Those of Leo LaPorte (1967), Perry Roehl (1967), Albert Matter (1967), and Paul Schenk (1969) are just a few of the many excellent examples that prove the importance of tidal flats as depositories for carbonate sediments.

FACTORS CONTROLLING THE DISTRIBUTION OF MARINE STROMATOLITES

Yesterday's heresy is today's dogma, and the bandwagon effect makes adherents to the new tidal-flat paradigm easier to muster than proof. Indulgent carping aside, how can we be sure that ancient stromatolite-binding algal mats had the same environmental distribution as those of today? To answer this, we must know what controls the distribution of Recent algal mats (Fig. 1).

The upper limit of mats in arid regions, such as the Trucial coast, is the top of the intertidal zone (Kendall & Skipwith 1968). In semiarid regions, such as Shark Bay, supratidal mats occur but are thin and poorly developed (Hoffman et al., in press). In areas of high rainfall, such as the Bahamas and south Florida, where the supratidal zone is moist or even flooded for days at a time, mats are prolific even in areas well above the limits of normal tides (Black 1933; Monty 1967; Shinn et al. 1969). The upper limit of Recent mats is apparently controlled by desiccation and there seems to be no reason why this limit should have been different in the past. Alfred Fischer (1965), however, suggested that algal mats may not have grown subaerially in the

Proterozoic because of the lethal effects of ultraviolet radiation in the absence of an atmospheric oxygen screen. This idea is countered by the hundreds of meters of stromatolites riddled with spar-filled shrinkage pores in the early Proterozoic Utsingi Formation (Hoffman 1968) of the Great Slave Lake area in northern Canada. These rocks are identical to Fischer's (1964) own algal-mat *loferites* of the alpine Triassic and are generally considered to be the result of algal-mat growth in an environment of semi-continuous desiccation (Shinn 1968). Therefore, I doubt that ultraviolet radiation prevented algal mats from colonizing Proterozoic tidal flats.

It is more difficult to explain the lower limit of Recent algal mats. In the Bahamas, sediment-binding mats are best developed in the uppermost parts of the intertidal zone. Peter Garrett (1970) showed that algae are prevented from establishing mats by the grazing and burrowing of cerithid gastropods and worms. Similarly, in the Trucial coast and the less saline parts of Shark Bay, the lower limit of mats coincides with the upper limit of algae-eating gastropods. Consequently, the mats are restricted to the upper parts of the intertidal zone, beyond the desiccation-tolerance of the gastropods. Only in

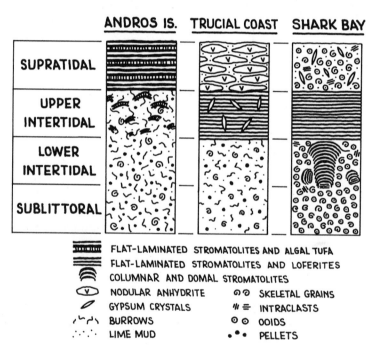

Fig. 1. Shoaling-upward sequences of carbonate sediment produced by progradation of Recent tidal flats in Andros island, Bahamas (Shinn et al. 1969); Abu Dhabi, Trucial coast (Kendall & Skipwith 1968); and Hamelin Pool, Shark Bay (Hoffman et al. in press).

the hypersaline bayheads, where salinity is too high for gastropods, do the Shark Bay mats extend through the lower intertidal zone, where they form the high-relief columnar stromatolites unique in the Recent. As Garrett points out, the great abundance of such stromatolites in the Proterozoic may be explained by the absence of grazing and burrowing animals, and consequently the greater prevalence of lower intertidal algal mats.

The importance of sublittoral stromatolites is difficult to evaluate because recent sublittoral environments are much less well known than tidal flats. A few ancient stromatolites may well have grown many meters below sea level on the foreslopes of carbonate platforms (Playford & Cockbain 1969; Hoffman 1972). Their deep-water origin is inferred from their location relative to an over-all platform-submarine slope-basin paleogeographic reconstruction, or because they are repeatedly interbedded with turbidites and other relatively deep-water sediments. Such stromatolites are clearly the exception not the rule, but I expect their numbers will increase.

STROMATOLITE DIAGENESIS AND THE DESTRUCTION OF PARTICULATE TEXTURE

How does one distinguish between fossil fresh-water algal tufa and stromatolites formed by sediment-binding algal mats? Where stromatolites are found encased in noncarbonate sediment, growth by in situ precipitation must be inferred. Where the surrounding sediment is calcareous, the choice is more difficult. The problem is that particulate textures, indicative of sediment-binding, are relatively rare in ancient stromatolites. This prompted Logan to begin a detailed study of the early diagenesis of the Shark Bay stromatolites, a preliminary report of which was presented at the Conference on Stromatolites at the Johns Hopkins University (Ginsburg 1967). The Shark Bay structures are composed of oöids, foram tests and other sediment particles washed onto the tidal flats from the adjacent shallow submarine platform. They differ from the unlithified stromatolites of Florida and the Bahamas in being penecontemporaneously cemented by cryptocrystalline aragonite, precipitated inorganically between the sediment particles. The particles themselves are progressively altered to cryptocrystalline aragonite as a result of algal boring. The end product is a rock that retains little evidence of its original particulate texture, especially after the inevitable conversion of the cryptocrystalline aragonite to calcite. Thus, the lack of particulate texture in ancient stromatolites may not be sufficient to rule out a sediment-binding origin.

STROMATOLITE MORPHOLOGY AND PALEOCURRENTS

Although direct evidence of sediment-binding can seldom be found in the texture of ancient stromatolites, indirect evidence is provided by their morphology. In an attempt to find stromatolites analogous to the linear *spur-*

and-groove structure of modern reefs (Shinn 1963), an idea suggested decades before by Winifred Goldring (1938), I measured the orientation of elongate domal stromatolites in the Proterozoic of Great Slave Lake and found a regionally consistent pattern (Hoffman 1967). Their elongation is everywhere parallel to the direction of sediment movement as determined from the foresets of ripple marks and crossbeds. Furthermore, domal stromatolites tended to be oversteepened and columnar ones inclined in the direction from which the sediment came.

I was delighted when Logan showed me the same relations in Shark Bay. Because of the refraction of waves, sediment movement onto the tidal flats is perpendicular to the shoreline. The elongation of domal stromatolites is also perpendicular to the shoreline. The domal stromalites tend to be oversteepened and columnar ones inclined toward the off-shore source of sediment. This is because the rate of sediment accretion is greatest on the up-current side of the stromatolite. The remarkable similarity between the Proterozoic and recent stromatolites convinces me that they both are formed by the same sediment-binding mechanism. In addition, the orientation and polarity of ancient shorelines can be easily determined from systematic measurements of stromatolite morphology.

STROMATOLITE LAMINATIONS AND THEIR TEMPORAL SIGNIFICANCE

Alan McGugan (1967) and Giorgio Pannella (Pannella et al. 1968) have recently measured rhythmic variations in lamination thickness of ancient stromatolites, following a suggestion of Keith Runcorn (1966) that stromatolites, like certain solitary corals, might reveal the number of days in synodic months of the past. These studies were made in the hope of being able to plot changes in the speed of the earth's rotation through geologic time, data of great importance to cosmology. The interpretations will be valid only if, as has been proved for modern corals, stromatolite laminations represent daily growth increments. To corroborate this, we must turn again to Recent stromatolites.

The growth rates of Recent stromatolites are eaily measured by scattering insoluble pigment onto the growing surface and later counting the number of laminations added above the pigmented horizon during succeeding days. (That the laminations would not be produced in the absence of living algae can be demonstrated by locally killing the algal mat with a toxic solution and marking as before.) The measurements made by Claude Monty (1967) in the Bahamas and Conrad Gebelein (1969) in Bermuda indicated that sublittoral stromatolites do have daily growth increments. Monty warned, however, that intertidal stromatolites are often temporarily buried by excessive sediment influx and no laminations are added for several days. The picture was further complicated when it was found that intertidal stromatolites in south Florida grow at the rate of two laminations each day (Gebelein & Hoffman 1968), one with each tide, and that during rainstorms, earlier laminations are eroded and lost from the record. To add a final note of caution, the Shark Bay stromatolites grow much more slowly, only a few laminations being added each year. From this, I judge attempts to determine the length of the synodic month from nonskeletal structures such as stromatolites to be overly optimistic.

CRYPTALGAL STRUCTURES THAT LACK LAMINATIONS

Closely associated with stromatolites in many ancient carbonate sequences are beds and discrete growth structures morphologically like stromatolites but without laminations. Among such rocks are Alfred Fischer's (1964) *loferites*, which are riddled with spar-filled voids, and Jim Aitken's (1967) *thrombolites*, which have an irregular clotted texture. Both are thought to be *cryptalgal*, Aitken's useful term for rocks, including stromatolites, in which an algal origin is inferred but cannot be proved. Aitken believes, on the basis of facies analysis of lower Paleozoic rocks, that thrombolites develop in sublittoral environments but there are as yet no recent analogs to substantiate this. Unlaminated cryptalgal sediments similar to loferites occur extensively in the upper intertidal zone of Shark Bay, in contrast to the laminated stromatolites of the lower intertidal. The unlaminated sediments form only where the algal mat is dominated by nonfilamentous species, whereas the stromatolites are produced by filamentous mats (Hoffman et al. 1969). This vindicates Fischer's astute observation that ancient loferites generally lack filament moulds and might therefore be the product of coccoid algae.

STROMATOLITES AND THE PROBLEM OF SELECTIVE DOLOMITIZATION

In outcrops of ancient stromatolites, the laminations weather in relief because of alternations in grain size or composition. Many stromatolites consist of millimeter-thick laminations of dolomite alternating with calcite, as revealed by staining or color differentiation of weathered surfaces (dolomite brown, calcite blue-grey). Some stromatolites also have vertical or radial threads of dolomite that resemble tufts of algal filaments in recent stromatolites. Commonly, the dolomite laminations have carbonaceous residues absent in the calcite. The highly selective distribution of the dolomite suggests a primary or penecontemporaneous origin, but no Recent stromatolites have been found to contain such thin dolomite laminations. Most Recent stromatolites have alternating laminations rich and poor in organic matter, mainly gelatinous sheath material secreted around the algal cells during growth of the mat. Comparison of ancient and Recent stromatolites indicates that the distribution of dolomite in the ancient corresponds exactly to the distribution of organic matter in the Recent. Conrad Gebelein (Gebelein & Hoffman 1969) has measured the magnesium content of algal sheath material in both laboratory cultures and natural Recent stromatolites, measurements that show selective concentration of magnesium in the organic matter. Gebelein claims that sufficient magnesium is contained in a 2 mm thick algal mat to produce a 1 mm thick layer of dolomite. The state of magnesium in the organic matter is not known but it is not in the form of dolomite. Perhaps dolomitization occurs long after deposition, the ultimate distribution of dolomite merely following the primary distribution of magnesium-rich organic matter. Perhaps this sort of dolomite after organic matter is the secret behind many of the laminated and mottled dolomitic rocks over which we have argued so long.

STROMATOLITES OF INORGANIC ORIGIN

Something that haunts geologists working on ancient stromatolites is the thought that they might not be biogenic at all. That this is a real possibility is best illustrated by the notorious case of the "algal pisolites" in the Capitan Reef complex of the Permian Basin, west Texas. The early history of the case is familiar—the pisolites were first thought, by Rudolf Ruedemann (1929), to be fossil coralline algae, but Harlan Johnson (1942), the American authority on coralline algae, believed them to be a form of stromatolite produced by blue-green algae. No less an authority, Julius Pia (1940), was among the handful that believed the pisolites to be inorganic. The authors of the classic book on the reef complex (Newell et al. 1953) noted their microscopic resemblance to inorganic travertine but ended up by siding with the majority opinion in favor of the algae.

The issue was unexpectedly resolved by Carroll Thomas (1968) and Robert Dunham (1969), who independently showed that the pisolites had a completely different Recent analog—pisolitic *caliche*, or *calcrete*, a type of stony calcareous soil formed inorganically in the vadose zone. Vadose pisolites form in situ and can be distinguished by having fitted polygonal shapes, reverse graded bedding, downward elongation, perched inclusions and laminations that coat pisolites, interstices and crosscutting fractures at the same time. Other examples of calcrete mistaken for stromatolites will doubtless be found, and we shall all rest easier when more criteria are available for indentifying fossil calcrete, particularly the nonpisolitic varieties.

BIOSTRATIGRAPHIC ZONATION OF THE PROTEROZOIC BASED ON STROMATOLITES

The emphasis of this review reflects the North American bias in studying stromatolites primarily for their value as environmental indicators and placing great importance on the lessons learned from Recent stromatolites. There has always been the hope, however, that from the study of ancient stromatolites alone might come a worldwide biostratigraphic zonation of the Proterozoic. Individual stromatolite beds can be traced for hundreds of kilometers in many sedimentary basins but, until the last decade, attempts at inter-basin correlation proved fruitless. This was not surprising considering that Recent stromatolites are built by heterogeneous algal communities, and their morphology seems environmentally rather than biologically controlled. Furthermore, Proterozoic algal microfossils are little different from extant genera, indicating extreme evolutionary conservatism (Schopf 1968). If the algae have changed little since the early Proterozoic, how can stromatolites be expected to show evolutionary trends?

In the face of all logic, an attempt was made to subdivide the Proterozoic in the Soviet Union and, to even their own astonishment, a degree of success was almost immediately achieved. A handful of specialists, working simultaneously in widely separated regions, arrived simultaneously at a four-fold zonation of the middle and late Proterozoic (for a review of this work published in English, see Raaben 1969). The zonation is based on changes in the shape, bordering, mode of branching, and internal laminations of columnar stromatolites. Tentative support for the Soviet zonation has been reported from Australia (Glaessner et al. 1969), India (Valdiya 1969), and the United States (Cloud & Semikhatov 1969). This raises the exciting possibility of world-wide correlations of Proterozoic rocks, albeit not in the same detail as in younger rocks.

Perhaps the most intriguing and certainly the best studied of columnar stromatolites are those belonging to the form-genus *Conophyton*, the subject of a treatise by three of the principal Soviet specialists (Komar et al. 1965). This stromatolite differs from all others in having acute conical laminations and a distinctive axial zone. Columns of *Conophyton* have an extraordinary size range—as narrow as 5 mm to giants of 10 m in diameter. Zonation of the Soviet Proterozoic is based in part on changes in the thickness and continuity of laminations within *Conophyton* columns. But the most intriguing fact is that this stromatolite has never been found in post-Proterozoic rocks and is so far unknown in the Recent. Small *Conophyton* columns occur in beds overlying intertidal stromatolites near the tops of shoaling-upward cyclic sequences in the early Proterozoic Rocknest Formation (Hoffman 1972) of nothern Canada, and I interpret these to have formed in stagnant supratidal marshes. The great synoptic relief of the larger *Conophyton* columns, however, makes a sublittoral environment more plausible (Donaldson & Taylor 1972).

The zonation is, as yet, far from perfect and many geologists, myself included, remain to be convinced that it has worldwide validity. The mode of branching of stromatolite columns, once important in establishing the zonation, has been found to vary widely within individual bioherms and has been downgraded by some specialists in favor of internal microtexture. Certain early Proterozoic columnar stromatolites in Canada closely resemble the branching and bordering of stromatolites diagnostic of the late Proterozoic in the USSR (Hofmann 1969; Hoffman 1970). Above all, the infinite variability of stromatolites makes objective definition of different forms exceedingly difficult. What is certain, however, is that the biostratigraphic potential of stromatolites can be tested only in ancient rocks and the USSR has every right to the claim of being the only nation where the extraordinary diversity of Proterozoic stromatolites has been adequately documented.

The possibility that stromatolites have evolved through time points up the principal inadequacy of the Recent—that it is not an exact replica of the past. Its value is in that it is the only geologic interval alive. Only in the Recent can a stromatolite be related directly to the organisms that built it and the environment in which those organisms live. The interpretation of ancient stromatolites will continue to depend on discoveries of Recent analogs.

REFERENCES

Aitken, J. D. 1967. Classification and environmental significance of cryptalagal limestones and dolomites, with illustrations from the Cambrian and Ordovician of southwestern Alberta. *Jour. Sedimentary Petrology 37*, 1163–78.

Black, M. 1933. The algal sediments of Andros Island, Bahamas. *Phil. Trans. Roy. Soc. London, ser. B, 222*, 165–92.

Blackwelder, E. 1913. Origin of the Bighorn Dolomite of Wyoming. *Geol. Soc. America Bull. 24*, 607–24.

Bradley, W. H. 1929. Algae reefs and oolites of the Green River formation, in *United States Geol. Surv., Prof. Paper 154*, 225–66.

Cloud, P. E., Jr., & Semikhatov, M. A. 1969. Proterozoic stromatolite zonation. *Am. Jour. Sci. 267*, 1017–61.

Donaldson, J. A., & Taylor, A. H. (in press). Conical-columnar stromatolites and the subtidal environment (abstract). *Soc. Econ. Paleontologists and Mineralogists*, Ann. Meeting, Denver, 1972.

Dunham, R. J. 1969. Vadose pisolite in the Capitan Reef (Permian), New Mexico and Texas, in Friedman, G. M. (ed.), *Depositional environments in carbonate rocks*, Soc. Econ. Paleontologists and Mineralogists, Spec. Publ. No. 14, pp. 182–91.

Eardley, A. J. 1938. Sediments of Great Salt Lake, Utah. *American Assoc. Petroleum Geol. Bull. 22*, 1305–1411.

Fischer, A. G. 1964. The Lofer cyclothems of the Alpine Triassic, in Merriam, D. F. (ed.), *Symposium on cyclic sedimentation, State Geol. Surv. Kansas, Bull. 169*, 107–50.

———. 1965. Fossils, early life and atmosphere history. *Proc. Nat. Acad. Sci. United States 53*, 1205.

Garrett, P. 1970. Phanerozoic stromatolites: noncompetitive ecologic restriction by grazing and burrowing animals. *Science 169*, 171–73.

Gebelein, C. D. 1969. Distribution, morphology and accretion rate of Recent subtidal algal stromatolites, Bermuda. *Jour. Sedimentary Petrology 39*, 46–69.

Gebelein, C. D., & Hoffman, P. 1968. Intertidal stromatolites and associated facies from Lake Ingraham, Cape Sable, Florida (abstract), *in Abstracts for 1968, Geol. Soc. America, Spec. Paper No. 121*, p. 109.

———. 1971. Algal origin of dolomite in interlaminated limestone-dolomite sedimentary rocks, *in* Bricker, O.P. (ed.), *Carbonate cements*, The Johns Hopkins University Studies in Geology No. 19. Baltimore: The Johns Hopkins Press, pp. 319–27.

Ginsburg, R. N. 1955. Recent stromatolitic sediments from south Florida (abstract). *Jour. Paleo. 29*, 723.

———. 1967. Stromatolites. *Science 157*, 339–40.

Glaessner, M. F., Preiss, W. V., & Walter, M. R. 1969. Precambrian columnar stromatolites in Australia: morphological and stratigraphic analysis. *Science 164*, 1056–58.

Goldring, W. 1938. Algal barrier reefs in the Lower Ozarkian of New York. *New York State Mus. Bull., No. 315*, 75 pp.

Hall, J. D. 1833. *Cryptozoon (proliferum)* N. G. (and sp.). *New York State Mus. Ann. Rept. No. 36*, pl. 6 and explanation.

Hoffmann, H. J. 1969. Attributes of stromatolites. *Geol. Surv. Canada, Paper 69–39*, 58 pp.

Hoffman, P. 1967. Algal stromatolites: use in stratigraphic correlation and paleocurrent determination. *Science 157*, 1043–45.

———. 1968. Stratigraphy of the Lower Proterozoic Great Slave Supergroup, East Arm of Great Slave Lake, District of Mackenzie. *Geol. Surv. Canada, Paper 68–42*, 93 pp.

———. 1970. Study of the Epworth Group, Coppermine River area, District of Mackenzie, in *Geol. Surv. Canada, Paper 70–1, Pt. A*, pp. 144–49.

———. In press. Proterozoic stromatolites of cyclic shelf, mounded shelf-edge and turbidite off-shelf facies, northwestern Canadian Shield (abstract). *Soc. Econ. Paleontologists and Mineralogists*, Ann. Meet, Denver, 1972.

Hoffman, P., Logan, B. W., & Gebelein, C. D., In press. Algal mats, cryptalgal fabrics and structures, Hamelin Pool, Shark Bay, in *Amer. Assoc. Petroleum Geologists, Memoir*.

Johnson, J. H. 1942. Permian lime-secreting algae from the Guadalupe Mountains, New Mexico. *Geol. Soc. America, Bull. 53*, 195–226.

Kalkowski, E. 1908. *Oolith und Stromatolith im norddeutschen Bundsandstein. Zeitschr. Deutsch. Geol. Ges. 60*, 68–125.

Kendall, C.St.C., & Skipwith, Sir P. d'E., Bt. 1968. Recent algal mats of a Persian Gulf Lagoon. *Jour. Sedimentary Petrology 38*, 1040–58.

Komar, V. A., Raaben, M. E., & Semikhatov, M. A. 1965. *Konofitony rifeya SSSR i ikh stratigraficheskoe znachenie. Akad. nauk SSSR, Geol. Inst. 131*, 73 pp.

Laporte, L. F. 1967. Carbonate deposition near mean sea level and resultant facies mosaic. Manlius Formation (Lower Devonian) of New York State. *Amer. Assoc. Petroleum Geologists, Bull. 51*, 73–101.

Logan, B. W. 1961. *Cryptozoon* and associated stromatolites from the Recent of Shark Bay, Western Australia. *Jour. Geology 69*, 517–33.

Logan, B. W., Rezak, R., & Ginsburg, R. N. 1964. Classification and environmental significance of algal stromatolites. *Jour. Geology 72*, 68–83.

Matter, A. 1967. Tidal flat deposits in the Ordovician of Western Maryland. *Jour. Sedimentary Petrology 37*, 601–9.

McGugan, A. 1967. Possible use of algal stromatolite rhythms in geochronology (abstract), in *Abstracts for 1967, Geol. Soc. Amer., Spec. Paper, No. 115*, p. 145.

Monty, C. L. V. 1967. Distribution and structure of recent stromatolitic algal mats, eastern Andros Island, Bahamas. *Ann. Soc. Geol. Belgique 90*, 55–100.

Newell, N. D., Rigby, J. K., Fischer, A. G., Whiteman, A. J., Hickox, J. E., & Bradley, J. S. 1953. *The Permian Reef Complex of the Guadalupe Mountains Region, Texas and New Mexico.* San Francisco: W. H. Freeman and Co., 236 pp.

Pannella, G., MacClintock, C., & Thompson, M. N. 1968. Paleontological evidence of variations in length of synodic months since late Cambrian. *Science 192*, 792–96.

Pia, J. V. 1940. *Vorläufige Ubersicht der kalkalgen des Perms von Nordamerika. Akad. Wiss. Wien., Math. Naturwiss. Kl., Anz. 9.*

Playford, P. W., & Cockbain, A. E. 1969. Algal stromatolites: deepwater forms in the Devonian of Western Australia. *Science 165*, 1008–10.

Raaben, M. E. 1969. Columnar stromatolites and Late Precambrian stratigraphy. *Am Jour. Sci. 267*, 1–18.

Roddy, H. J. 1915. Concretions in streams formed by the agency of bluegreen algae and related plants. *Proc. American Philos. Soc. 54*, 246–58.

Ruedemann, R. 1929. Coralline algae, Guadalupe Mountains. *Amer. Assoc. Petroleum Geologists Bull. 13*, 1079–80.

Runcorn, S. K. 1966. Corals as paleontological clocks. *Scientific American 215*, 26–33.

Schenk, P. E. 1969. Carbonate-sulfate-redbed facies and cyclic sedimentation of the Windsorian Stage (Middle Carboniferous), Maritime Provinces. *Canadian Jour. Earth Sci. 6*, 1037–66.

Schopf, J. W. 1968. Microflora of the Bitter Springs Formation, Late Precambrian, central Australia. *Jour. Paleontology 42*, 651–88.

Shinn, E. 1963. Spur and groove formation on the Florida reef tract. *Jour. Sedimentary Petrology 33*, 291–303.

_____. 1968. Practical significance of birdseye structures in carbonate rocks. *Jour. Sedimentary Petrology 38,* 215-23.

Shinn, E., Lloyd, R. M., & Ginsburg, R. N. 1969. Anatomy of a modern carbonate tidal-flat. *Jour. Sedimentary Petrology 39,* 1202-28.

Tebbutt, G. E., Conley, C. D., & Boyd, D. W. 1965. Lithogenesis of a distinctive carbonate rock fabric. *Wyoming Geol. Surv., Contributions to Geology 4,* 1-13.

Thomas, C. M. 1968. Vadose pisolites in the Guadalupe and Apache Mountains, west Texas, in *Permian Basin Section, Soc. Econ. Paleontologists and Mineralogists, Symposium and Guidebook Publ. 68-11,* 32-35.

Valdiya, K. S. 1969. Stromatolites of the Lesser Himalayan carbonate formations and the Vindhyans. *Jour. Geol. Soc. India 10,* 1-25.

Walcott, C. D. 1914. Pre-Cambrian Algonkian algal flora. *Smithsonian Misc. Collections 64,* 74-156.

Weber van Bosse, A., & Foslie, M. 1904. The corallinaceae of the *Siboga* Expedition. *Siboga Expedition Repts. 61,* 110. Amsterdam, Leiden: Brill.

Wieland, G. R. 1914. Further notes on Ozarkian seaweeds and oolites. *Am. Mus. Nat. History Bull. 33,* 237-60.

THE JOHNS HOPKINS UNIVERSITY PRESS

This book was composed in Press Roman text and Univers display type by
Jones Composition Company, Inc. It was printed by Universal Lithographers, Inc.
on S. D. Warren's 60-lb. Sebago, in a text shade, regular finish and bound by
L. H. Jenkins, Inc. in Columbia Bayside linen.

Library of Congress Cataloging in Publication Data

Main entry under title:

Evolving concepts in sedimentology.

(The Johns Hopkins University studies in geology, no. 21)
At head of title: For Francis Pettijohn, teacher and geologist.
Essays first read at a conference at Johns Hopkins University in January 1971.
Includes bibliographies.
CONTENTS: R. G. Walker. Mopping up the turbidite mess.—H. P. Eugster. Experi-
mental geochemistry and the sedimentary environment: Van't Hoff's study of marine
evaporites.—K. J. Hsu. The odyssey of geosyncline. [etc.]
1. Rocks. Sedimentary—Congresses. 2. Sedimentation and deposition—Congresses. I. Gins-
burg, Robert N., ed. II. Pettijohn, Francis John, 1904- III. Johns Hopkins University.
IV. Series.
QE471.E9   551.3'04   72-4016
ISBN 0-8018-1444-8